Engines
of Mischief

REACTING TO THE PAST is an award-winning series of immersive role-playing games that actively engage students in their own learning. Students assume the roles of historical characters and practice critical thinking, primary source analysis, and argument, both written and spoken. Reacting games are flexible enough to be used across the curriculum, from first-year general education classes and discussion sections of lecture classes to capstone experiences, intersession courses, and honors programs.

Reacting to the Past was originally developed under the auspices of Barnard College and is sustained by the Reacting Consortium of colleges and universities. The Consortium hosts a regular series of conferences and events to support faculty and administrators.

NOTE TO INSTRUCTORS: Before beginning the game you must download the game materials, including an instructor's manual containing a detailed schedule of class sessions, role sheets for students, and handouts.

To download these essential resources, visit https://reactingconsortium.org/games, click on the page for this title, then click "Game Materials."

Engines of Mischief

TECHNOLOGY, REBELLION, AND THE INDUSTRIAL REVOLUTION IN ENGLAND, 1817–1818

LOUISE BLAKENEY WILLIAMS,
BRENDAN PALLA, AND
MEGAN SQUIRE

BARNARD

The University of North Carolina Press

Chapel Hill

© 2024 Reacting Consortium, Inc.
All rights reserved
Manufactured in the United States of America

Cover art: *Leader of the Luddites*, 1812. Hand-colored etching. Artist unknown. "General," or Ned, Ludd was a legendary weaver who led his followers to break the new machines and burn down factories during the early Industrial Revolution. In this image he is disguised in women's clothing.

ISBN 978-1-4696-8354-6 (pbk.: alk. paper)
ISBN 978-1-4696-8355-3 (epub)
ISBN 978-1-4696-8356-0 (pdf)

Contents

List of Illustrations / vii

1. INTRODUCTION / 1

Brief Overview of the Game / 1

Prologue: A Walk to the Pub / 2

Basic Features of Reacting to the Past / 4

   Game Setup / 4

   Game Play / 5

   Game Requirements / 5

Controversy / 6

Counterfactuals / 7

A Note on Monetary Value / 7

2. HISTORICAL BACKGROUND / 9

Setting / 9

Chronology / 9

The Industrial Revolution, Manchester, and Responses to Economic Change, 1812–1818 / 10

The Industrial Revolution / 12

Condition of England Debate / 15

Economic Distress / 19

Working-Class Protest and Government Response / 19

Demand for Political Reform / 21

The Middle Classes and New Economic Theories / 25

Challenges to the New Economic Theory / 25

Conclusion / 26

3. THE GAME / 27
    Major Issues for Debate / 27
    Rules and Procedures / 28
        Objectives and Victory Conditions / 28
        Status Points / 28
        Winning / 28
    Outline of the Game / 29
        Location and Time / 29
        Activities for Each Session / 29
        Accounting / 31
        Possible Other Events / 31
        General Election / 31
    Example Schedule of Sessions / 32
    The Many Functions of the Game Master / 33
    The Game Master News Service / 33
    Assignments and Grading / 33

4. ROLES AND FACTIONS / 35
    Social Classes / 35
    Factions / 36
    Indeterminacy / 36
        Secrets / 36
    Roles / 36
        Aristocracy / 36
        Middle Classes / 38
        Working Classes / 42

5. CORE TEXTS / 45
    Adam Smith, *The Wealth of Nations* / 45
    Robert Owen, *A New View of Society* / 61
    Thomas Malthus, *An Essay on the Principle of Population* / 73
    Proposals to Help Handloom Weavers / 78
    Report on Child Labor / 89
    On Taxation / 100
        On the Corn Laws / 100
        On the Income Tax / 105
        On Other Taxes / 107
        Testimony to Parliament about Taxation / 111
    On the Poor Laws / 113
        Testimony to Parliament about the Poor Laws / 113
        Summary of a Report to Parliament about the Poor Laws / 116
        Thomas Courtenay on the Poor Laws / 118
    On Political Reform / 119
    Reports of Working-Class Protest / 125
    The Riot Act and Combination Acts / 128

Acknowledgments / 133
Notes / 135
Selected Bibliography / 137

Illustrations

FIGURES

2.1 Spinning and handloom weaving in a cottage / 11

2.2 Mule spinning in a factory / 11

2.3 Power loom weaving in a factory / 11

2.4 Cotton factories, Union Street, Manchester / 14

2.5 Luddite rioters / 21

4.1 Weavers' cottages, Manchester / 44

MAP

2.1 Industry in Britain, 1715–1815 / 8

Engines
of Mischief

1 Introduction

BRIEF OVERVIEW OF THE GAME

Engines of Mischief is set in Manchester, England, in 1817 and 1818. This was a period of wage crisis, class conflict, and rapid technological change during the early years of the Industrial Revolution. The game that you will play over the next few weeks offers a snapshot of life among the different classes of people living and working in Manchester at this time.

The players are drawn from all classes of society, from lords to laborers and everything in between. Textile workers, craftsmen, merchants, and aristocrats are faced with different choices about how to live and prosper during the period of great technological, economic, and social change that we now call the Industrial Revolution. Each character in the game has their own motivations and will approach political, economic, and social issues differently.

This game provides a platform for deep discussion of the complexities of the Industrial Revolution by engaging students in serious reading of key historical texts and prompting debate about the following issues:

- Industrialization: the impact of machine labor and the factory system.
- Labor exploitation, capitalism, and socialism.
- The balance between benevolence and self-interest.
- The nature of class and status.
- The role of the government in the economy.
- Political power and powerlessness.
- Why revolutions do, or do not, occur in different places and times.

PROLOGUE: A WALK TO THE PUB

It is a cold dreary day as you walk toward Manchester's Market Place, another wet week with bad prospects for work. The gray clouds and drizzle match your mood. As a skilled handloom weaver you once earned twenty-one shillings each week for producing cloth in your home. But over the last decade the merchants who put out the raw cotton for spinning into thread and weaving into cloth have continually cut your wages until now they stand at a paltry ten shillings per week. You are single and would like to marry, but how do you support a wife and children on ten shillings per week, let alone maintain your equipment? At first, they blamed the long wars with Napoleon for the drop in trade, but it is now 1817 and two years since Wellington defeated the emperor at Waterloo and exiled him to St. Helena. The British navy is the most powerful in the world and protects British trade all over the globe—what excuse is there now for impoverishing talented, hardworking English laborers?

As you near the square you get one answer as the light wind brings you smoke from the city's paper mills and iron foundries. A haze settles over the edge of the city where steam engines now power a variety of manufactures, including power looms that work day and night. You remember back in the 1790s when you rejoiced at the inventions that allowed children of weavers to work at home with their families, running spinning machines with twelve spindles. Letting the women and children spin freed you to make more cloth. But then jennies were invented with thousands of spindles run by waterpower and production had to move out into factories. You sniff at the thought of the coarse textiles woven by machines that are so different from the "fine" cloth you and other weavers produce working in your homes. But those machines have started to drive down the wages of the skilled home production workers. You know something of the factories as well, because your sister worked in one from the age of ten. She tells stories of clouds of dust and lint that made it difficult to breathe, and noise from the machines that made it impossible to hear. Quiet conversation between workers was impossible, even if the overseers had given any time for breaks. She arrived at 6 a.m. and worked until 7 p.m., six days a week, snatching any breakfast she could from her pockets while working, and taking only thirty minutes for lunch—although that time was frequently spent cleaning the machines. Since then, your sister married a handloom weaver and now she works helping her husband by spinning the yarn he needs to produce his woven textiles. Her work is still hard, but she prefers to be at home where she can tend her small children. Better that than to be at the mercy of the machines.

It is Monday, a day you traditionally take to clean and repair tools and to recover from the festivities of Sunday. On Tuesday you'll start to work hard to meet your weekly quota of cloth but even then, you set your own pace. This is a day to go to town to hear the latest gossip. While you can read and write a little, you like to hear the news at the Weavers' Arms public house. Frequently the publican reads from the latest newspaper, or traveling journeymen tell stories about events in the larger world. The pub is a place to complain with other weavers, and to talk about how life could be improved. The songs you sing there frequently have a hard edge. Some of the tunes brought by travelers in the last decade have praised a certain General Ludd and warned merchants and aristocracy to be careful. You particularly like the song "The Hand-Loom Weavers' Lament" (to the tune of "A Hunting We Will Go"):

> You gentlemen and tradesmen that ride about at will,
> Look down on these poor people. It's enough to make you crill.
> Look down on these poor people, as you ride up and down
> I think there is a God above will bring your pride quite down.

Chorus
You tyrants of England! Your race may soon be run.
You may be brought unto account for what you've sorely done.
You pull down our wages, shamefully to tell.
You go into the markets and say you cannot sell.
And when that we do ask you when these bad times will mend,
You quickly give an answer, "When the wars are at an end."
When we look on our poor children, it grieves our hearts full sore.
Their clothing it is worn to rags, while we can get no more.
With little in their bellies, they to work must go,
Whilst yours do dress as manky as monkeys in a show.
You go to church on Sundays. I'm sure it's naught but pride.
There can be no religion where humanity's thrown aside.
If there be a place in heaven, as there is in the Exchange,
Our poor souls must not come near there. Like lost sheep they must range.
With the choicest of strong dainties, your table's overspread
With good ale and strong brandy, to make your faces red.
You call'd a set of visitors—It is your whole delight—
And you lay your heads together to make our faces white.
You say that Bonyparty he's been the spoil of all,
And that we have got reason to pray for his downfall.
Well, Bonyparty's dead and gone, and it is plainly shown
That we have bigger tyrants in Boneys of our own.
And now, my lads, for to conclude, it's time to make an end,
Let's see if we can form a plan that these bad times may mend.
Then give us our old prices, as we have had before,
And we can live in happiness and rub off the old score.[1]

The singing is in good fun, but the weavers are listening more and more to the sharp tone of the words. Dare you go farther than talking with your neighbors? But you must be careful, it is illegal for workers to form any kind of combination (i.e., a union), and should you decide to take action, the magistrate can easily turn to the Riot Act to declare any meeting illegal. When the Riot Act is read, participants can be arrested, or worse.

You don't fear hard work, in fact you are working harder than ever before, and yet you find it difficult to keep yourself living, much less prospering. If things don't mend soon you may have to try something else or go somewhere else. But where can you turn for help? Parliament has recently abolished all the old laws that protected guilds and apprenticeship. What arguments could you use to convince the local aristocracy of the need to protect skilled workers? They seem kindly enough toward the lower classes, but with their servants and carriages and house parties and fox hunts they clearly have no sense of what life is like for you. You hope to convince the merchants that they need to raise your pay, but with rumors that they can hire Irish immigrants on the cheap, how can you convince them of your worth? You have heard that more and more children are being employed in factories. This frightens you on many levels: what about the welfare of the children, what about the quality of the work they produce, and what about your own paycheck? What words can you find to convince the merchants and the aristocrats of the importance of protecting quality English jobs? More important, can you convince them that they should enact laws and standards to protect the welfare of the traditional English family? Can you turn to the local vicar for help? Perhaps the skilled craftsmen who have traveled and read will help you find the right words.

All you know as you approach the pub is that things can't stay as they are. But you are wary of what you say, and to whom; who knows where the magistrate has ears? You have trained to produce traditional-style cloth since you were a boy. How can you save your way of life? How can you provide for a family? How can you make the aristocrats and merchants care? How can you defend your craft and your way of life in the face of change?

BASIC FEATURES OF REACTING TO THE PAST

This is a historical role-playing game. After a few preparatory lectures, the game begins and the students are in charge. Set in moments of heightened historical tension, it places you in the role of a person from the period. By reading the game book and your individual role sheet, you will find out more about your objectives, worldview, allies, and opponents. You must then attempt to achieve victory through formal speeches, informal debate, negotiations, and conspiracy. Outcomes sometimes differ from actual history; a debriefing session sets the record straight. What follows is an outline of what you will encounter in Reacting and what you will be expected to do.

Game Setup

Your instructor will spend some time before the beginning of the game helping you to understand the historical context for the game. During the setup period, you will use several different kinds of material:

- The game book (what you are reading now), which includes historical information, rules and elements of the game, and essential historical documents.
- A role sheet, which provides a short biography of the historical person you will model in the game as well as that person's ideology, objectives, responsibilities, and resources. Some roles are based on historical figures. Others are "composites," which draw elements from a number of individuals. You will receive your role sheet from your instructor.

Familiarize yourself with the documents before the game begins and return to them once you are *in role*. They contain information and arguments that will be useful as the game unfolds. A second reading while in role will deepen your understanding and alter your perspective. Once the game is in motion, your perspectives may change. Some ideas may begin to look quite different. Those who have carefully read the materials and who know the rules of the game will invariably do better than those who rely on general impressions and uncertain memories.

Game Play

Once the game begins, class sessions are presided over by students. In most cases, a single student serves as some sort of presiding officer. The instructor then becomes the GM (the "game master" or "game manager") and takes a seat in the back of the room. Though they do not lead the class sessions, GMs may do any of the following:

- Pass notes
- Announce important events
- Redirect proceedings that have gone off track

Instructors are, of course, available for consultations before and after game sessions. Although they will not let you in on any of the secrets of the game, they can be invaluable in terms of sharpening your arguments or finding key historical resources.

The presiding officer is expected to observe basic standards of fairness, but as a fail-safe device, most games employ the "podium rule," which allows a student who has not been recognized to approach the podium and wait for a chance to speak. Once at the podium, the student has the floor and must be heard.

Role sheets contain private, secret information that you must guard. Exercise caution when discussing your role with others. Your role sheet probably identifies likely allies, but even they may not always be trustworthy. However, keeping your own counsel and saying nothing to anyone is not an option. In order to achieve your objectives, you *must* speak with others. You will never muster the voting strength to prevail without allies. Collaboration and coalition building are at the heart of every game.

Some games feature strong alliances called *factions*. As a counterbalance, these games include roles called *indeterminates*. They operate outside of the established factions, and while some are entirely neutral, most possess their own idiosyncratic objectives. If you are in a faction, cultivating indeterminates is in your interest, since they can be persuaded to support your position. If you are lucky enough to have drawn the role of an indeterminate, you should be pleased; you will likely play a pivotal role in the outcome of the game.

Game Requirements

Students playing Reacting games practice persuasive writing, public speaking, critical thinking, teamwork, negotiation, problem solving, collaboration, adapting to changing circumstances, and working under pressure to meet deadlines. Your instructor will explain the specific requirements for your class. In general, though, a Reacting game asks you to perform three distinct activities:

Reading and writing. What you read can often be put to immediate use, and what you write is meant to persuade others to act the way you want them to. The reading load may vary slightly from role to role; the writing requirement depends on your particular course. Papers are often policy statements, but they can also be autobiographies, battle plans, newspaper articles, poems, or after-game reflections. Papers often provide the foundation for the speeches delivered in class. They also help to familiarize you with the issues, which should allow you to ask good questions.

Public speaking and debate. In the course of a game, almost everyone is expected to deliver at least one formal speech from the podium (the length of the game and the size of the class will determine the number of speeches). Debate follows. It can be impromptu, raucous, and fast paced. At some point, discussions must lead to action, which often means proposing, debating, and passing a variety of resolutions. GMs may require that students deliver their papers from memory when at the podium, or they may insist that students begin to wean themselves from dependency on written notes as the game progresses.

Wherever the game imaginatively puts you, it will surely not put you in the present. Accordingly, the colloquialisms and familiarities of today's college life are out of place. Never open your speech with a salutation like "Hi guys" when something like "Fellow citizens!" would be more appropriate.

Always seek allies to back your points when you are speaking at the podium. Do your best to have at least one supporter second your proposal, come to your defense, or admonish inattentive members of the body. Note-passing and side conversations, while common occurrences, will likely spoil the effect of your speech; so you and your supporters should insist upon order before such behavior becomes too disruptive. Ask the presiding officer to assist you. Appeal to the GM as a last resort.

Strategizing. Communication among students is an essential feature of Reacting games. You will likely find yourself writing emails, texting, attending out-of-class meetings, or gathering for meals. The purpose of frequent communication is to lay out a strategy for achieving your objectives, thwarting your opponents, and hatching plots. When communicating with fellow students in or out of class, always assume that they are speaking to you in role. If you want to talk about the "real world," make that clear.

CONTROVERSY

Most Reacting games take place at moments of conflict in the past and therefore are likely to address difficult, even painful, issues that we continue to grapple with today. Consequently, this game may contain controversial subject matter. You may need to represent ideas with which you personally disagree or that you even find repugnant. When speaking about these ideas, make it clear that you are speaking in role. Furthermore, if other people say things that offend you, recognize that they too are playing roles. If you decide to respond to them, do so using the voice of your role and make this clear. If these efforts are insufficient, or the ideas associated with your particular role seem potentially overwhelming, talk to your GM.

When playing your role, rely on your role sheet and the other game materials rather than drawing upon caricature or stereotype. Do not use racial and ethnic slurs even if they are historically appropriate. If you are concerned about the potential for cultural appropriation or the use of demeaning language in your game, talk to your GM.

Amid the plotting, debating, and voting, always remember that this is an immersive role-playing game. Other players may resist your efforts, attack your ideas, and even betray a confidence. They take these actions because they are playing their roles. If you become concerned about the potential for game-based conflict to bleed out into the real world, take a step back and reflect on the situation. If your concerns persist, talk to your GM.

COUNTERFACTUALS

Most of the characters in the game were people who lived at the time. Some historical information can be found about many of them, but they often are composites of several contemporaries. In addition, the number of players of each social class in the game does not reflect historical reality; there were far more members of the working classes than the middle or upper classes.

The game only uses shillings as currency. Historically, the English had a system of pence, shillings, pounds, with twelve pence in a shilling and twenty shillings in a pound. There also were other coins like florins, crowns, and guineas. Students may find these referred to in the readings, but to simplify matters the game uses shillings alone.

Voting in the game attempts to reflect historical reality. This is why the male members of the aristocracy are the only characters who can vote in Town Hall or for Parliament. In the game, aristocrats can vote on laws that would apply to all of Britain, such as an income tax, the Corn Laws, or the Poor Laws, when in reality they would only be able to change the laws for the town of Manchester. But the premise of our game is that our aristocrats have been asked by the prime minister of England to break a deadlock in Parliament in London. Parliament has been debating bills related to Manchester and the other new industrial towns but does not have enough knowledge of the new economy to decide definitively about them. They want the advice of the leaders of Manchester society. This is why, if the majority of aristocrats of Manchester vote a certain way on a proposed law, Parliament will pass that law for all of England. In short, the Manchester aristocrats will tip the balance in Parliament. But the aristocrats want to hear from all the people of Manchester who will be impacted by any changes made by new parliamentary laws. So, any Manchester resident is allowed to make arguments in Town Hall to convince the aristocrats which way to vote.

A NOTE ON MONETARY VALUE

To get a sense of what it meant for a worker to be paid 10 shillings (s) per week or a machine to cost 200s in our game, it is useful to consult a contemporary guidebook on domestic economy. This book claims that 21s a week (about 55 pounds [£] per year) was the minimum income for a working-class family of a husband, wife, and three children to live respectably. With this wage they could pay for rent, coal, and candles but would eat mostly bread (24 pounds per week), a little fish and meat, and some beer, but no tea, sugar, or other luxuries.[2] The richest family described in the book had an income of £5,000 per year (1,923s a week) or ten times that of the poor worker. They could afford twenty-two servants, ten horses, a coach (four-wheeled carriage) and two other two-wheeled carriages, four cows and a garden with a gardener, medicine and medical attention, education, and entertainment. They spent almost twice as much in tea and sugar alone than the poor working-class family spent on all their expenses, 36s per week. They also consumed 175s per week in meat and fish, and 140s per week in beer, wine, and alcohol.[3] But this family still was part of the middle classes. Most aristocrats had incomes of more than £10,000 per year (as does Mr. Darcy in Jane Austen's 1813 novel *Pride and Prejudice*), and a few were worth hundreds of thousands of pounds.

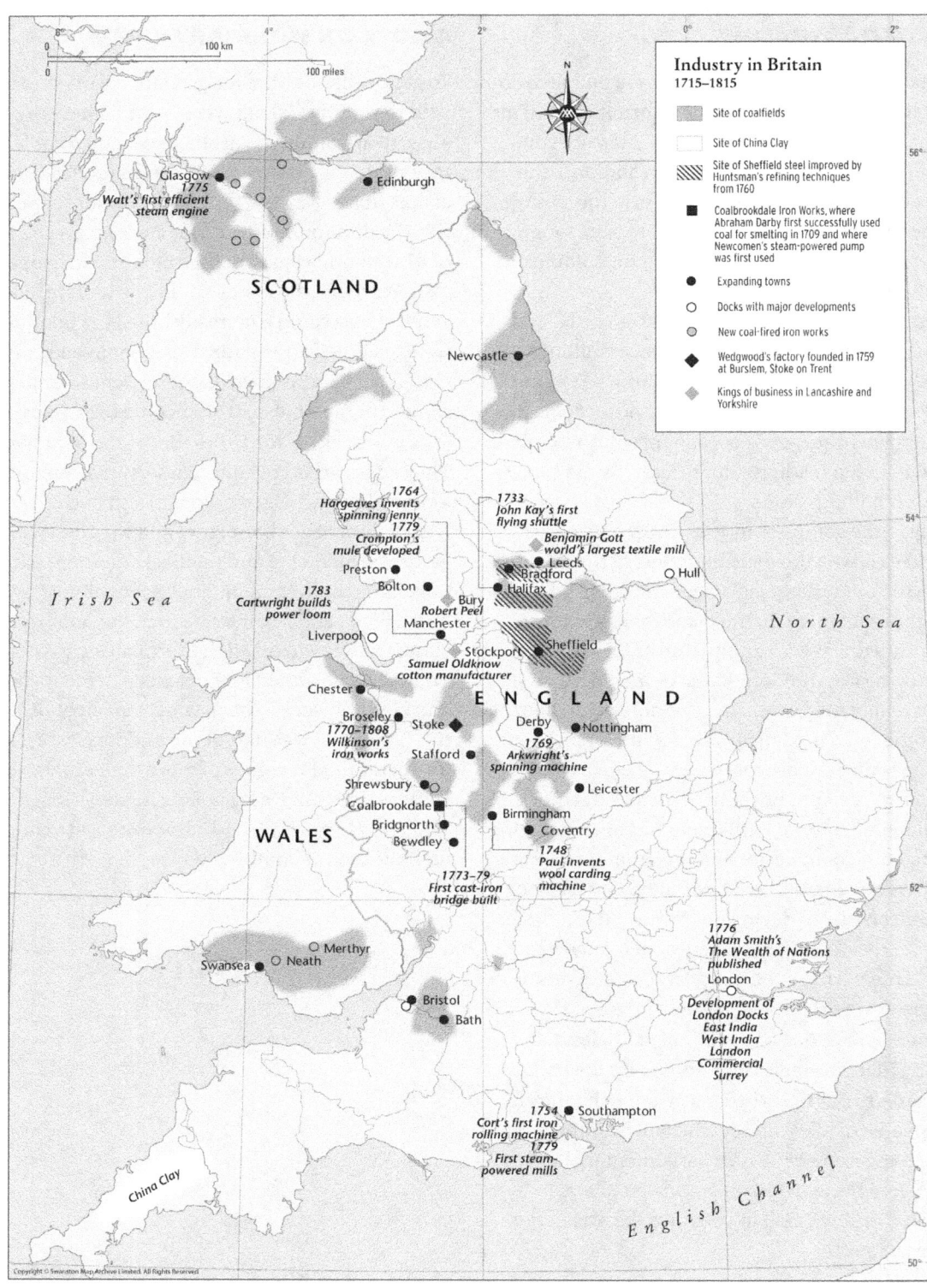

Courtesy Axiom Maps Limited.

2

Historical Background

SETTING

This game is set in Manchester, England, in 1817 and 1818. Manchester is located in the area of England south of Scotland, northwest of London, and across the Irish Sea from Ireland. It is located in the county of Lancashire, which borders on Yorkshire to the northeast, the East Midlands to the southeast, and the West Midlands to the south. Map 2.1 shows some of the nearby towns.

CHRONOLOGY

1712 First steam engine invented by Thomas Newcomen.

1733 Flying shuttle invented by John Kay.

1739 Methodist preachers (John Wesley and others) begin their mission to the poor.

1760 George III becomes monarch.

1764 Spinning jenny invented by James Hargreaves.

1769 James Watt's steam engine granted a British patent.

1769 James Watt invents the measurement of "horsepower."

1769 Water frame invented by Richard Arkwright.

1771 First cotton spinning mill opened by Richard Arkwright using his water frame.

1774 Spinning mule invented by Samuel Crompton.

1776 American Revolution begins.

1783 American Revolution ends.

1784 Iron made in a puddling furnace invented by Henry Cort.

1785 Power loom invented by Edmund Cartwright.

1787 Committee for the Abolition of the Slave Trade forms.

1789 French Revolution begins.

1790 Wool combing machine invented by Edmund Cartwright.

1792 Coal gas used to light the home of William Murdock (James Watt's assistant).

1793 Cotton gin invented by Eli Whitney.
1793 Britain goes to war with France.
1798 Rebellion in Ireland.
1799 Combination Act becomes law.
1801 Act of Union creates United Kingdom of Great Britain and Ireland.
1803 Cotton textiles become Britain's biggest export, overtaking wool.
1804 Napoleon Bonaparte becomes emperor of France.
1804 First steam locomotive railway journey takes place.
1807 Great Britain abolishes the slave trade.
1811 Regency Act makes George IV prince regent.
1811 First large-scale Luddite riot takes place in Nottingham.
1812 Law passes making the destruction of industrial machines punishable by death.
1813 Fourteen Luddites hanged in Manchester.
1815 Defeat of Napoleon and end of wars with France.
1815 Corn Laws introduced to protect British agriculture.
1816 Income tax abolished.
1816 Steam engine locomotive patented by George Stephenson.

THE INDUSTRIAL REVOLUTION, MANCHESTER, AND RESPONSES TO ECONOMIC CHANGE, 1812–1818

"What Art was to the ancient world, Science is to the modern: the distinctive faculty. In the minds of men the useful has succeeded to the beautiful . . . a Lancashire village has expanded into a mighty region of factories and warehouses. Yet rightly understood, Manchester is as great a human exploit as Athens."[1] Benjamin Disraeli made this comparison when he wrote about the manufacturing power of Manchester in 1844. At that time, 70 percent of Britain's textile industry was located in this town. Of the 300,000 residents of Manchester, half were involved in the cotton textile industry, most as factory workers. Manchester's factories consumed annually a billion pounds of cotton, largely imported from the United States or India.

One hundred years earlier, Manchester, like the rest of England, was a predominantly agricultural community with a population of only 40,000 people. There were no factories for making cloth, or for making anything else. Cloth was produced from raw materials such as wool or flax, materials that were grown locally. Weaving families spun thread and wove cloth in their own homes for mostly domestic distribution using the traditional methods of home production: hand carders, a spinning wheel, and a narrow frame loom.

Figures 2.1, 2.2, and 2.3 symbolize the change in the weaving industry. Figure 2.1 shows a vignette of the home production process for making cloth. The woman uses a hand-operated spinning wheel to convert raw cotton into thread, which the man then weaves by hand into cloth. They work together in the attic of their own cottage, where they also live. Figures 2.2 and 2.3 show spinning and weaving in a factory in Manchester several decades later. This factory had five floors, and each floor held dozens of machines, each producing thousands of pounds of cloth per day. New inventions allowed one semiskilled worker to replace dozens of skilled workers. The spinners and weavers merely tended

FIGURE 2.1 Before the Industrial Revolution textiles were made using hand-operated spinning wheels and looms. From Henry Edward Tidmarsh, *The Age of Industry: Spinning and Hand Loom Weaving, 1893–94.* © Manchester Art Gallery/Bridgeman Images.

FIGURE 2.2 During the Industrial Revolution machines in factories spun raw cotton into thread. *Mule Spinning*, engraving, in Edward Baines, *History of the Cotton Manufacture in Great Britain* (London: H. Fisher, R. Fisher, and P. Jackson, 1835), 210. Image courtesy HathiTrust.

FIGURE 2.3 In factories, power looms also wove thread into textiles. *Power Loom Weaving*, engraving, in Edward Baines, *History of the Cotton Manufacture in Great Britain* (London: H. Fisher, R. Fisher, and P. Jackson, 1835), 239. Image courtesy HathiTrust.

the machines, rather than made the cloth themselves. By 1830, the home production method of making cloth was becoming a relic of the past.

THE INDUSTRIAL REVOLUTION

These changes in production are part of what is called the Industrial Revolution. This "revolution" involved a fundamental shift from the manufacturing of goods at home and by hand, to the use of machines in factories using nonanimal power, such as water or steam. It was not, however, revolutionary in the sense that it happened overnight. Industrialization started in England in the 1760s. Its effects were widely felt by the 1820s. By 1850, industry employed more people than agriculture.

The first industry to mechanize was cotton textiles. The application of steam power to machines eventually made it possible to manufacture all sorts of other goods and transport them around the world on railways and steam ships. When this occurred, the Industrial Revolution expanded exponentially.

There were many causes of the Industrial Revolution, but one important one was an increased demand for cotton textiles. In fact, the period from approximately 1788–1803 was called "the golden age" of cotton cloth production.[2] There was a huge demand for cloth in England because improvements in agriculture in the first half of the century resulted in increased food production and lower food prices. More food meant more people lived to childbearing years. This, in turn, led to a huge expansion of the population and the ability of the working classes to afford goods beyond mere subsistence, like cotton clothing. Raw cotton also was easy to come by because of the British Empire. Slavery was firmly in place in the Americas, and it provided a seemingly inexhaustible supply of raw cotton, even after the American Revolution. Ship transport was safe enough and reliable enough to be counted on to bring cotton to England and deliver finished textiles to an expanding world market. In fact, cotton imports to England tripled during this time.

William Radcliffe, a weaver and merchant, described how these changes impacted the hand-produced cotton textile industry near Manchester:

> An increasing demand for every fabric the loom could produce, put all hands in request of every age and description. . . the old loom-shops being insufficient, every lumber-room, even old barns, cart-houses, and outbuildings of any description were repaired, windows broke through the old blank walls, and all fitted up for loom-shops. This source of making room being at length exhausted, new weavers' cottages with loom-shops rose up in every direction; all immediately filled, and when in full work the weekly circulation of money as the price of labour only rose to five times the amount ever before experienced in this sub-division, every family bringing home weekly 40, 60, 80, 100, or even 120 shillings per week!!![3]

With such a demand, merchants eventually realized that changes had to be made to cotton production. Since the early eighteenth century the so-called **putting-out system**, or **cottage industry**, was most common.

In the **putting-out system,** or **cottage industry**, merchants rode on horseback from house to house delivering raw cotton fibers. The cloth was made by hand in the artisans' homes. Then the merchant came by again to the same house to pick up the finished cloth and pay the spinners and weavers a weekly wage.

But it was soon clear that this was not the most efficient way to make textiles to meet the growing market. This was why merchants latched onto new inventions that promised to increase production significantly. These inventions helped speed up the various steps in producing cotton cloth. They made an extensive division of labor possible and engendered a factory system using machines.

Some of the most important new inventions included the *cotton gin*, which automated the tedious work of separating the cotton seeds from the fiber. Use of the gin resulted in large increases in the amount of raw material imported from cotton-producing nations. The *spinning jenny* improved upon the centuries-old spinning wheel by allowing one spinner to use a "Great Wheel" to spin thread onto eight spindles instead of one. The invention of the *flying shuttle* improved the 2,000-year-old handloom technology by allowing the shuttle to be thrown back and forth using one hand only, thus speeding up production.

Still, the jenny and flying shuttle were inventions that could be used with *home production* of cloth. Both of these inventions increased productivity of weavers but did not fundamentally change the nature of a weaver's work. In 1800, the spinning jenny and a handloom with a flying shuttle were both common equipment in weaving families' homes, and weaving could still be considered a family enterprise. Children were still put to work carding (straightening) the fibers with hand carders, women could still spin the thread on a wheel or jenny, and men still did the weaving with a narrow frame handloom. At the beginning of the century, home cloth production methods were still viable, but in just a few short years, the handwriting was on the wall—new inventions would decisively change how cloth was produced.

First, the invention and diffusion of the *carding engine* allowed the raw cotton fibers to be combed and aligned much faster than by hand carding with paddles. This removed the bottleneck between the deseeding and spinning processes. The *spinning mule*, an improvement on the spinning jenny, could spin 400 spindles with just a single worker. The primary job of this one worker was simply to ensure no broken threads. The *water frame*, so called because it used water wheel for power, could spin thread many times stronger than the thread produced by a spinning wheel or spinning jenny. Finally, a water wheel, and then a steam engine, was able to drive mechanical parts of a loom, performing the same motions formerly done by a weaver, and the *power loom* was invented. Steam engines were powered by huge amounts of coal, used to heat water into steam. One contemporary author, writing in 1810, noted that one "steam-engine, of the power of forty horses, consumes about . . . 11,000 lbs. weight of coals in twenty-four hours."[4] But these engines were preferred to water wheels because they did not need flowing water as a source of power and could be located anywhere in the country.

None of these inventions could fit into a home, so merchants opened factories to house and run the machines. The machines also rendered skilled labor unnecessary. In the factories workers were primarily engaged in unskilled tasks such as supervising the machines, picking up loose fibers from machinery (scavenging), or tying knots in broken threads (piecing). Many factory owners preferred to employ young women and children in these jobs since they could fit more easily under the machines and because their labor cost less than adult male labor.

Figure 2.4 shows that some factories could be very large. One French traveler who visited a cotton manufactory in Lanark, Scotland, around 1810 "saw four stone buildings, 150 feet front each, four stories high of twenty windows, and several other buildings" that employed "2500 workmen, mostly children, who work from six o'clock in the morning till seven o'clock in the evening."[5] The entire process of making cotton by machine impressed him:

> It is impossible to see without astonishment these endless flakes of cotton, as light as snow, and as white, ever pouring from the carding-machine, then seized by the teeth of innumerable wheels and cylinders, and stretched into threads, flowing like a rapid stream, and lost in the *tourbillon* of spindles. The eye of a child or of a woman, watches over the blind mechanism, directing the motions of her whirling battalion, rallying disordered and broken threads, and repairing unforeseen accidents. The shuttle likewise, untouched, shoots to and fro by an invisible force; and the weaver, no longer cramped upon his uneasy seat, but merely overlooking his self moving looms, produces forty-eight yards of cloth in a day, instead of four or five yards.[6]

FIGURE 2.4 Early factories in Manchester could be converted warehouses. *Cotton Factories, Union Street, Manchester*, engraving, in Edward Baines, *History of the Cotton Manufacture in Great Britain* (London: H. Fisher, R. Fisher, and P. Jackson, 1835), 395. Image courtesy HathiTrust.

Not only could these machines in factories produce huge amounts of textiles with far less time and effort, but some of those textiles also could be of a better quality than ones made by hand. Another traveler, who visited a factory in 1816 that employed 500 people and wove cotton cloth in two rooms each with 125 power looms, was especially impressed by the quality of machine-made cotton cloth. He noted that "cottons wove by the power looms fetch a much better price at market, than those wove in the usual manner," because they were "of a better texture than what is woven by a common loom."[7] On asking the partner of a cotton manufactory in Glasgow about this, he was told "that cottons wove in this manner are better than those made in the usual way," because of "the mathematical regularity with which the threads are placed."[8]

In 1817 the Industrial Revolution was well underway in Britain. But it was not nearly complete. While the spinning of thread was mechanized fairly rapidly, steam-powered weaving machines were still problematic. They may have made better "coarse" cloth, but they were not yet sophisticated enough to produce high-quality "fine" cloth. As a result, steam engines were used at the time primarily to make textiles from heavy thread, while handweavers still were necessary for more delicate cloth. This meant that a number of different business organizations could be found side by side for decades. Some weavers and spinners still worked in their own cottages on equipment they owned, while others left home to work in large establishments containing multiple handlooms or jennies owned by their employers. A few businesses combined steam power spinning machines with handweaving looms. And all of this existed at the same time as the first of the fully steam-powered factories were opened, which usually (although not always) were integrated in the sense that they had equipment for carding, spinning, and weaving under one roof. The transition to industry was slow, but steam-powered factories were clearly the future of cloth production.

Along with factories came industrial cities. A contemporary, writing in 1823, claimed that it was because of the influence of cotton manufacturing that "the town of Manchester has, from an unimportant provincial town, become the second in extent and population in England, and Liverpool has become in opulence, magnitude, elegance and commerce, the second Seaport in Europe."[9] He explained how this happened: "The origin of a Manufacturing town is this: a Manufactory is established, a number of labourers and artizans are collected—these have wants which must be supplied by the Corn Dealer, the Butcher, the Builder, the Shopkeeper—the latter when added to the Colony have themselves need of the Draper, the Grocer, &c. Fresh multitudes of every various trade and business, whether conducive to the wants or luxury of the inhabitants, are superadded, and thus is the Manufacturing town formed."[10]

CONDITION OF ENGLAND DEBATE

As this description of the origins of Manchester indicates, many contemporaries welcomed the Industrial Revolution and were optimistic about its benefits. Some historians agree that the Industrial Revolution had positive consequences. It clearly provided more opportunities for a wide variety of people. The middle classes, such as manufacturers, traders, and bankers, had new ways to make wealth. Jobs were provided for millions of workers. In addition, goods were cheaper to buy. Transportation was much easier, especially with the expanding railway system and steam ships. Cities provided excitement and entertainment.

However, other contemporaries and later historians noted some serious problems that accompanied industrialization. Work in factories could be difficult and dangerous, as there was no government regulation of working conditions. The working day was long, often sixteen hours. Factories were overheated, cotton filaments irritated workers' lungs, and accidents and injuries occurred too frequently. Workers had to leave the comfort of their homes for soulless establishments where strangers told them what to do and often imposed strict discipline for inattentiveness or insubordination. Rather than working at one's own pace at home, workers had to adhere strictly to a time clock. Work conditions were especially bad for children, who could be employed as young as five years old.

Living conditions and sanitation were noticeably poor in the new cities as well. Because cities grew up so quickly, there was an inadequate supply of housing and little to no sanitation. Workers were crowded into filthy slums. As a result, child mortality was appalling; 60 percent of children in Manchester died before reaching the age of five.

Poet Robert Southey, who visited Manchester in 1808, was especially concerned about the impact of industrialization on children, who, as he put it,

> are deprived in childhood of all instruction and all enjoyment; of the sports in which childhood

instinctively indulges, of fresh air by day and of natural sleep by night. Their health physical and moral is alike destroyed; they die of diseases induced by unremitting task work, by confinement in the impure atmosphere of crowded rooms, by the particles of metallic or vegetable dust which they are continually inhaling; or they live to grow up without decency, without comfort, and without hope, without morals, without religion, and without shame, and bring forth slaves like themselves to tread in the same path of misery.[11]

Southey also commented on the living conditions of workers: "The dwellings of the labouring manufacturers are in narrow streets and lanes, blocked up from light and air. . . . Here in Manchester a great proportion of the poor lodge in cellars, damp and dark, where every kind of filth is suffered to accumulate, because no exertions of domestic care can ever make such homes decent. These places are so many hotbeds of infection; and the poor in large towns are rarely or never without an infectious fever among them, a plague of their own."[12]

Southey was a poet and friend of many of the leading artists of the Romantic school. Most of those authors and artists also were troubled by industrialization. William Blake, for example, wrote of "dark Satanic Mills," and declared that "a machine is not a man nor a work of art," but instead is "destructive of humanity and art."[13]

To Romantic artists nature was particularly important because it was God's creation and thus a reflection of the divine. They worried that human industry was destroying God's great work. If one contemporary description of Manchester in 1818 is to be believed, the Romantics did have much to fear. According this author, in Manchester the "water is sometimes so damaged by dye-houses, and other works, erected upon rivers, as to be rendered not wholesome to the cattle, and destructive to fish. The heat necessary for the business of printing debilitates the strongest constitutions.—Damps from obstructed water; pestilential air from crowded rooms;—effluvia from acids and different preparations;—down from cotton; all operate as pestilences to the human constitution."[14]

Many Romantic artists also were concerned about the impact of industrialization on the working class. Lord Byron was particularly upset by the plight of handloom weavers. Percy Bysshe Shelley worried that the exploitation of workers by business owners had created a class "Fiercely thirsting to exchange / Blood for blood—and wrong for wrong."[15]

Perhaps the most chilling representation of the potentially negative consequences of industrialization came from the pen of Shelley's wife, Mary Wollstonecraft Shelley. In her 1819 novel *Frankenstein*, Shelley speculated about the disastrous results of the science that gave humans the power to meddle with nature. In this novel, scientist Dr. Victor Frankenstein invents a brilliant technique to defeat the natural process of mortality and bring the dead alive. The result is a hideous "monster" with tremendous powers. When Frankenstein rejects his creation, the monster seeks revenge, and it is the creation, not the creator, who eventually gains complete control. As the monster declares to Frankenstein, "I can make you so wretched that the light of day will be hateful to you. You are my creator, but I am your master;—obey!"[16] It could have been the machines invented to improve industrial production, or the members of working class who were made miserable by early industrial conditions, that Mary Shelley had in mind here.

While many people suffered from the transition to industrialization, those deeply attached to the older, traditional economy felt the change most severely. This was especially the case for handloom weavers who wished to continue producing cloth in their homes. As late as 1835 a parliamentary committee was formed to investigate the condition of these weavers. It concluded that "the testimony of the deep distress and lamentable destitution pervading the handloom weavers in cotton, silk, linen and wool, is so general and so universally admitted, as scarcely to require analysing."[17]

Because they were competing with machines, the handloom weavers found that their wages declined rapidly. As one manufacturer testified to Parliament in 1834, "The condition of the weavers engaged in handloom weaving I consider to be very bad; it does not appear to me that if a man with his wife, and, we will say, two children, regularly employed in full work, at the present general prices, could make anything like a decent living."[18] A member of Parliament (MP) who testified the next year noted that

> a very great number of the weavers are unable to provide for themselves and their families a sufficiency of food of the plainest and cheapest kind; that they are clothed in rags, and indisposed on this account to go to any place of worship, or to send their children to the Sunday schools; that they have scarcely anything like furniture in their houses; that their beds and bedding are of the most wretched description, and that many of them sleep upon straw; that notwithstanding their want of food, clothing, furniture and bedding, they for the most part have full employment; that their labour is excessive, not unfrequently 16 hours a day; that this state of destitution and excessive labour induces them to drink ardent spirits to revive their drooping spirits and allay their sorrows, whereby their suffering is increased.[19]

E. P. Thompson notes in *The Making of the English Working Class* that this situation was made worse for the independent home weavers because their services were still needed in times when orders exceeded the capacity of factories to meet demand for finished cloth. This meant that they never retrained to another occupation and continued to put faith in the traditional economy.[20]

In addition, the handloom weavers were prevented from using collective bargaining, forming unions, or engaging in wage or price negotiations. Parliament passed a **Combination Act** in 1799 and again in 1800 that made trade unions and other working-class organizations illegal. This was to ensure that the needs of growing industry and business enterprises were not inconvenienced by worker strikes, especially during a time of war.

Under the **Combination Acts** it was against the law for workers or employers to organize, take action, or make agreements in order to raise wages, reduce working hours or the amount of work, or prevent employers from hiring whomever they wanted. Only friendly and provident societies were permitted.

But the decline in wages and living conditions was not the only complaint handweavers had about the new economic system. They also resented the reduction in the status and dignity of their work it engendered. They harbored a "memory of lost liberties and independence" and of the "reputation and respectability" that weavers once had under the medieval **guild system**.[21] Although guilds were effectively gone by 1817, they were not forgotten. In particular, skilled workers like handloom weavers remembered how guilds ensured workers' control over the production and sale of what they made.

In the **guild system** guild masters controlled all parts of the trade; they trained apprentices and journeymen in the secrets of the production process, and they determined who could make and sell various products, the quality of materials, and the ultimate price to be charged. They also took care of their members; if one became too ill to work or died and left a family, the guild would provide financial support.

Handweavers particularly were proud to be skilled and independent; they were in charge of their work and the product of their labor. This made them a step above the common worker and gave them respectability and status. This also made them very reluctant to take charity from others or rely on government assistance. Much of what they disliked about the new machines and factory system was the deskilling of labor. Any worker could do the work and no apprenticeship was required. Workers also lost almost all their independence and controlled neither their work conditions nor the final product of their labor.

The handloom weavers were not the only ones who regretted the erosion of a traditional way of life. Landowners who lived near the new industrial cities complained that no one seemed to care any more about the traditional occupation of farming on which their livelihood depended. As one contemporary noted, "Never inquire about the cultivation of land, or its produce, within ten or twelve miles of Manchester; the people know nothing about it: speak of spinning-jennies, and mules, and carding machines, they will talk for days with you."[22]

Landowners also complained about the increase in the numbers of the poor. The manufacturers encouraged settlers from elsewhere when they needed workers, but when business slowed, they fired them and left them with no other means of subsistence. These paupers had to rely on government assistance to say alive. This meant that Poor Rates increased, but it was the landowners, not the manufacturers, who had to pay these taxes.[23] The result, according to one contemporary, was that in the countryside around Manchester it was possible to "witness such appearance of poverty, exemplified in nakedness, dirtiness, and the different garbs which indicate distress. There are mendicants of all ages and sexes, but more particularly in the country villages; the exerted police of well-governed towns restrains these wanderers."[24]

The new factories also impacted landowners negatively because they drew workers away from agriculture and pushed up wages. This made it harder for farmers to profit and pay high rents to landowners. Many agricultural workers preferred working in factories, where both the wages and the work conditions were better. One observer at the time noted that "the advance of wages, and the preference given to the manufacturing employment, by laborers in general, where they may work by the piece, and under cover, have induced many to forsake the spade for the shuttle, and have embarrassed the farmers, by the scarcity of workmen, and of course advanced the price of labour."[25] He continued, "Who will work for 1s. 6d. or 2s. a day at a ditch, when he can get 3s. 6d. or 5s. a day in a cotton work, and be drunk four days out of seven?"[26] The attractiveness of factory work is supported by one contemporary writer who noted that "almost all the farmers, who have raised fortunes by agriculture, place their children in the manufacturing line" and that many of the better-off farmers have moved into trade. This was both because of "the enormous and immoderate wages to be obtained in the manufactories" and also because they were drawn by "the great wealth which has in many instances been so rapidly acquired by some . . . fortunate adventurers" in industry.[27]

These descriptions of workers who liked the new industry and preferred it to agricultural labor appears to contradict the deplorable descriptions of factory conditions given by contemporary observers. However, a recent historian argues that this was not the case. After consulting over 350 autobiographies of workers, Emma Griffin argues that while they disliked the work they did as children, as adults many factory workers appreciated the higher wages and regular employment.[28] They still disliked their lost independence, however, and the breakup of the family economy engendered by factory work. The fact that they no longer employed their children in their own homes and lost "an opportunity of bringing them up, and giving them good moral instruction" was deeply disturbing to many.[29]

We can conclude, therefore, that early industrialization was deeply uneven, even paradoxical, in its effects. This is perhaps best summed up by French thinker Alexis de Tocqueville, who visited Manchester in 1838. As he wrote in his diary, "From this foul drain, the greatest stream of human industry flows out to fertilise the world. From this filthy sewer pure gold flows. Here humanity attains its most complete development and its most brutish; here civilisation works its miracles, and here civilised man is turned back almost into a savage."[30]

ECONOMIC DISTRESS

The problems that the handloom weavers, and all members of the working class, faced in the early stages of industrialization in England were exacerbated by political developments and the continental wars that had a serious impact on the British economy.

For twenty-five years, including much of the first two decades of the nineteenth century, Britain was at war with France. The British declared war in 1793 after the French Revolution was well underway. The new French government threatened to spread the revolution to the rest of the monarchies of Europe and appeared to be able to do so. The wars intensified after Napoleon Bonaparte got control of the government in 1799 and declared himself emperor in 1804. These wars cost a great deal of money. To pay for them, the British government was forced to raise taxes and even impose an income tax. Neither the French nor the British were able to win a decisive victory; the French dominated on land, but the British commanded the seas. To end this stalemate, in 1806 Napoleon imposed the Continental System, which forbade his allies and satellite states from trading with the English. The British responded with Orders in Council that required all ships going to Europe to stop at a British port and pay customs duties. This seriously impacted trade and businesses in both Britain and on the Continent. On top of this, for two years after 1812 Britain was at war with *both* France and the United States.

Life in England during this time was difficult. Food prices and mortality rates were high. Even though both wars were over in 1815, the hardships continued. Rural hunger drove many country folk to cities. Increased populations in cities led to housing shortages, sanitation problems, and unrest. As bad as things were in England, they were worse in Ireland. Following the failed 1798 Irish Rebellion, Ireland officially was joined with Britain into the United Kingdom in 1801. This led to increased Irish immigration across the English Channel and into Liverpool and Manchester. The massive influx of Irish labor persisted through the first half of the nineteenth century and had the effect of lowering wages, especially in the most dangerous and lowest-paid occupations.

WORKING-CLASS PROTEST AND GOVERNMENT RESPONSE

In response to these economic problems, the working classes used a number of different, and often overlapping, tactics to improve their lives. All of them, however, relied on the understanding that individual laborers were powerless against the elite, and that their only strength was in their numbers. Collective action was most effective. The riot was a traditional form of collective action in the early modern period and was still used periodically around 1817. Riots were common, but they were not generally revolutionary or random. Rather, they were a form of negotiation. "Negotiation by riot" could pressure employers or the government to treat common people fairly and respect what workers saw as their legitimate rights.[31] Stores might be broken into, property smashed, or offenders attacked if the price of food was deemed too high or business owners did not respect guild practices.

A slightly better-organized, but not dissimilar tactic was that of the so-called **Luddites**, who were especially active in 1808, 1811–12, and in Lancashire during 1816–17 and who attempted to destroy the new machinery that was competing with hand production.

Luddites were workers in hand production who used an old guild practice of destroying property in order to extract concessions from employers. Claiming that they were working under the direction of a fictitious General Ludd, these loose confederations operated in secrecy, staging midnight raids on the factories and workshops of merchants and manufacturers who employed technology or practices they did not like. Their most common goal was to wreck the new machines that competed with them.

One contemporary song describes the actions of higher paid handworkers in the woolen branch of the cloth industry, known as croppers:

> And night by night when all is still
> And the moon is hid behind the hill,
> We forward march to do our will
>   With hatchet, pike, and gun!
> Oh, the cropper lads for me,
> The gallant lads for me,
> Who with lusty stroke
> The shear frames broke,
> The cropper lads for me![32]

The Luddites *twisted in* new members with secret oaths and ceremonies; they swore allegiances and pledged loyalty and silence as they leveled the factories and foul, obnoxious machines with "hatchet, pike, and gun." They sent threatening letters like the one below:

Mr Harvey

This is to inform you that if you do make any more two course Hole, you will have all your Frames broken and your Goods too, though you may think you have made your doom just I shall know how to break your frames, we will not suffer you to win the Trade will die first, if we cant do it just to night we will break them yet, and if we cant break them we can break something better and we will do it too in spite of the Devil

Remember Nedd Ludd[33]

Figure 2.5 shows the ferocity of the Luddites. Some Luddite songs insisted that the workers were not to blame because they were pushed beyond the breaking point. "Brave Ludd" in the song "General Ludd's Triumph" was "to measures of violence unused" and did not aim at vengeance, but "his suffering became so severe" because the new technology lowered prices and wages that he was forced to become "the grand Executioner." And so "These Engines of mischief were sentenced to die / By unanimous vote of the Trade."[34] Regardless of the cause, the organized violence of the lower classes frightened the government, which responded with harsh measures. All who could be captured were arrested and tried. Machine-breaking was made a capital crime in 1812, so some Luddites were executed. The lucky ones were transported to Australia.

Instead of, or in addition to, rioting or smashing machines, some workers organized themselves into proto-trade unions. Trade societies or associations in late eighteenth-century Britain regularly negotiated with employers, and often used strikes to pressure them to compromise. They were generally tolerated because the government recognized that the workers were trying to ensure fairness, and to protect their traditional skill and status, not overturn the established order. Some thought unions promoted industrial and political peace by helping to solve popular grievances.

Workers also recreated some of the protections of guilds in ways that later would be taken over by trade unions. They formed "friendly societies," or "box clubs" in local pubs, which would collect regular contributions in order to provide economic assistance in difficult times or upon death. In fact, one-third to one-half of the population of entire towns could belong to a friendly society.[35]

During and after the French Revolution, however, governments were less inclined to tolerate working-class organizations or violence because of their fear of the spread of revolution to Britain. This was the reason for the passage of the Combination Acts of 1799 and 1800. But these acts were only selectively enforced. Only a few people were prosecuted under them at first, and local governments were not required to enforce them unless a business owner asked them to do so, which they rarely did if workers did not strike. Governments feared that enforcing these laws might antagonize workers, who had legitimate grievances, and would encourage, rather than prevent, serious disturbances. The elites also did not want to prosecute business owners for combining to lower wages, which also was prohibited under these

FIGURE 2.5 Violence often resulted when Luddites organized to destroy the new machines. Hablot Knight (Phiz) Browne, *Luddite Rioters*, 1813. Private collection/Bridgeman Images.

laws. If they targeted workers too frequently the inequity of not punishing employers would become apparent and problematic.

Even if they were not always enforced, the Combination Acts did have an impact. Workers had to organize in secret. The laws made it clear that the aristocracy was willing to help employers at the expense of laborers. And as a result, the acts may have pushed workers to become more politically radical as a way to empower themselves in the face of elites who had joined together against them.[36]

DEMAND FOR POLITICAL REFORM

The spread of political radicalism among the working classes seriously alarmed the government. After the execution of King Louis XVI during the French Revolution there was a persistent fear among the upper classes in England that laboring Britons would attempt a revolution of their own. In the 1790s, in response to demands for reform, Prime Minister William Pitt created his own "Reign of Terror" by banning seditious publications and meetings of over forty people, arresting leaders of reform societies, and declaring it treason to incite contempt of constitution or of king. Spies were everywhere in Britain on the lookout for revolutionaries even as late as 1817. As one historian put it, the government "maintained extensive and sophisticated intelligence-gathering networks throughout the country, penetrating almost all aspects of society."[37]

While Pitt may have briefly suppressed demands for political reform, he was unable to end them entirely. When economic prosperity did not immediately return to Britain after the wars ended in 1815, some citizens blamed the government, and again began to call for reform.

On top of this, in 1815 and 1816 Parliament passed a number of laws that particularly angered the middle and working classes and suggested that the government was working only in the interests of the aristocracy. The first measure was the **Corn Laws**.

The **Corn Laws** required that when the price of domestic grain fell too low, either high tariffs would be imposed on cheap foreign grain, making it more expensive than British grain, or foreign grain would be prohibited from being imported into the country entirely. This was to ensure that high-priced British grain, sold by aristocrats and their tenant farmers, was purchased at home, rather than cheaper foreign grain.

The Corn Laws were intended to help the farmers and landowners make a profit because they kept the price for domestic grain sales high. This enabled aristocratic landowners to charge more in rent to the farmers who leased their land. But these laws had negative consequences for others. They increased the price of bread, which exacerbated the economic suffering of the poor. This, in turn, meant that middle-class business owners had to raise wages so their workers could afford food. And they went against the free trade ideas of economists like Adam Smith, which business owners increasingly supported.

In 1816, Parliament abolished the direct income tax, which had been imposed in 1799 to pay for the Napoleonic Wars. It was a progressive tax on income from property, businesses, trade, and offices. Those people who were wealthy paid 10 percent, those in the middling classes had a sliding scale of 1 percent to 10 percent, and those who were poor paid no tax at all. In other words, those least able to pay did not suffer.

When the French wars were over, this income tax was repealed under pressure from the aristocracy and wealthy members of society. But the government still needed money to cover its massive debts, so it had to raise many indirect taxes on goods to compensate. The result was that the poor had to pay more for things like sugar, beer, tea, soap, candles, clothing, bricks, and paper.

In the end, because of these changes, the poor ended up paying *more* in taxes after the war and the wealthy ended up paying *less*. In fact, one contemporary estimated that for a working man "at the very least, half of his income is abstracted from him by taxation."[38]

It is hardly surprising that in these conditions the middle and working classes began to pay more attention to radicals, like Henry Hunt and William Cobbett, who argued that the British government was an aristocratic oligarchy that had to be reformed to better represent the vast majority of the population. The qualifications for voting for, as well as sitting in, Parliament seemed to support this conclusion. The only people who could vote were men who were worth quite a bit of money. Even more income was required to be elected a member of Parliament. To vote, a man had to own freehold land that could be rented for 40 shillings per year. An income of £600 per year from freehold land was necessary to sit in Parliament.[39]

There were two types of electoral districts in nineteenth century Britain; county districts included the people of the countryside and boroughs were for the towns and cities. In county districts the electors were largely tenants of the local aristocrats, and because votes were taken in public without a secret ballot, tenants almost always voted for the candidate of their landlord's choice so as not to face eviction. Borough elections also were under the control of the aristocracy. The borough districts had been designated in the Middle Ages and not altered since. Therefore, they did not reflect recent changes in population. Some boroughs had so few electors that elections were not even contested. In these "pocket" or "rotten" boroughs the local aristocrat simply chose who would be the member of Parliament. In other boroughs the number of electors was so small that, if an election were held, an aristocrat could easily bribe the voters to choose his candidate. In other areas, such as the new industrial cities like Manchester, the huge population was not represented at all. The city of Manchester was in the county seat of Lancashire that had only two members of Parliament to represent over 100,000 people.

Changes within the British royal family seemed to support the argument that Britain's aristocratic government cared little for the vast majority of the population. The monarch was the top of the aristocratic hierarchy, but in 1817 the popularity of the monarchy was at an all-time low. This was a period known as the **Regency**.

The **Regency** refers to the period after 1811 when King George III was deemed unfit to reign due to mental illness (in actuality a disease called *porphyria*), so his son ruled in his stead as prince regent. In 1821 the prince officially took the throne as George IV and reigned until 1830. George IV's brother William IV reigned afterward, until he died in 1837, at which point William's niece Victoria ascended the throne. In some contexts, the term *Regency* is used to describe the entire transitional period between *Georgian* and *Victorian*, or roughly 1811 to 1837.

George IV as regent could not be more of a contrast to his father. George III was an economical family man with deep religious convictions, who eventually was greatly admired as the "father" of the nation. His eldest son, George IV, was an extravagant and self-indulgent spendthrift. Devoted to elaborate parties, gambling, wearing fancy clothes, and consuming rich food and large quantities of alcohol, he became fat and immensely indebted. He also ignored his wife for his many mistresses, on whom he also spent a great deal of money. On top of this, he was an extreme Tory, or conservative, who would tolerate no political change. He became a symbol of the selfishness and waste of aristocratic power and privilege.

With all this at work, it is hardly surprising that in his seminal work *The Making of the English Working Class*, E. P. Thompson calls the four-year period between 1815 and 1819 "the heroic age of popular Radicalism" in England.[40]

Radical ideas spread to the working class in many ways. National and local publications were either read aloud in a pub or read by the workers themselves. Richard Guest, writing in 1823, noted that the working class had recently become more literate and political than ever. He attributed it to the Sunday schools that were established starting in the 1780s, where workers "learned to read, write and cast accounts." He also noted "the countless publications dispersed over the country" that were cheap enough for working men to afford. For example, the radical *Cobbett's Weekly Political Register* sold for two pence. These publications helped laborers "reason and think for themselves," and become "Political Citizens" who were well informed about issues of war and peace, rising and falling wages, and what was going on in the national government.[41] Two of these cheap, especially labor-friendly London publications that had wide circulation in Manchester were the *Gorgon* and the *Black Dwarf*.

The *Gorgon* was radical enough to suggest that reform might even include granting women the right to vote. In the end, however, this was deemed impractical because the "whole *female gender*" are "in themselves *nothing*."[42] In fact, this was true legally at the time. Under English law in 1817, women had few rights. This was especially the case for **married women**.

Married women had few rights in Regency England, because under the "principle of coverture" the legal existence of a wife was incorporated into that of her husband. A married woman could not own property or make contracts in her own name. If she gained any property, it became that of her husband to do with as he wished. She could not get an education, make decisions about their children, or move without her husband's permission. Divorce was practically impossible.

The English philosopher Mary Wollstonecraft claimed in 1790 that these laws made women equivalent to slaves and so must be changed. She also thought women deserved the right to vote, higher education, and jobs. In the early nineteenth-century Wollstonecraft was distinctly unpopular among middle- and upper-class Britons. However, many working-class women were active in promoting political reform, initially for men, but possibly for themselves in the future. Some women attended, and were even allowed to vote at, meetings where radical political ideas were spread.[43]

Key figures who brought controversial political ideas to the working class included middle-class reformers like Major John Cartwright, William Cobbett, and Henry Hunt. They held large public meetings and gave impassioned speeches advocating for "radical" changes like universal male suffrage, a secret ballot, payment of members of Parliament, and the redistribution of seats to allow for more representation for new industrial cities like Manchester. Cartwright also started a network of private clubs that he named Hampden Clubs to educate all classes about the need for political change, and to coordinate reform activities across the country.

In Manchester in January and February 1817, speakers at some public meetings connected the economic plight of the suffering cotton textile workers with the need to extend the right to vote. It was decided that the prince regent himself had to be made aware of this through a petition. On 10 March 1817, a meeting of about 25,000 people was held in St. Peter's Field, Manchester, to see off 5,000 marchers who would carry the petition to the prince in London. They were mostly spinners and weavers, and each carried a blanket or coat to keep warm as they slept by the road along the way. But just as these "Blanketeers" set off on their journey, the Riot Act was read, and the meeting was broken up. The few men who were able to begin the march were attacked by the cavalry and arrested.

All of this working-class activity frightened some members of Parliament so much that they were convinced a revolution was imminent. In 1817 a secret committee of the House of Commons believed it had discovered that working-class radicals who advocated reforming Parliament were also hoping for "a total overthrow of all existing establishments" and the "division of the landed, and extinction of the funded property of the country."[44] The committee concluded that workers were planning a violent revolution that would create "general confusion, plunder, and bloodshed."[45]

This is not to say there was no revolutionary underground in early nineteenth-century Britain. A "Spencean Society" did unsuccessfully try to provoke rioting and a takeover of the government at a large meeting at Spa Fields in London in December 1816. And many trade societies were sympathetic to the aims of the revolutionary underground. But the extent of support for revolution was much smaller than the government supposed.

THE MIDDLE CLASSES AND NEW ECONOMIC THEORIES

It was not just the working classes who hoped to see a reform of Parliament. Many in the newly wealthy middle classes did as well, although they certainly did not advocate a violent revolution to accomplish or accompany reform. In fact, innovative business owners understood that economic problems had contributed to the revolutionary aspirations of workers, and they were confident that the new industrial mode of production would contribute to solving them. If only industrialists could have the right to vote and sit in Parliament along with the aristocracy, changes to the economy could be effected that would bring wealth to all people in the nation. They represented the new economy of industry and moveable wealth, or capital, in opposition to the aristocracy who were still clinging to the older economy of landownership and rent.

The middle classes and new factory owners hoped that the government could embrace the Enlightenment economic theories of thinkers like Adam Smith. They advocated laissez-faire, which included limited government interference in the economy, free trade, and a free market, to help the manufacturing interests in opposition to mercantilist government interventions like the Corn Laws favored by the landed aristocracy. If this was done wealth would "trickle" down to the poor, the entire nation would become rich and powerful, and all desire for revolution would end.

CHALLENGES TO THE NEW ECONOMIC THEORY

But not everyone was as optimistic as some businessmen that complete economic freedom was the answer to the new industrial economy. With laissez-faire went the elimination of government intervention to guarantee decent wages, working conditions, or job security for workers. Some people argued that this simply would make matters worse. Economic freedom, according to critics of laissez-faire, meant the freedom for business owners to pay their employees just barely enough to survive. As one citizen testified to Parliament as late as 1834, the "immense supply of mechanical labour has really given such a power to capital over the labour of the working classes that it is almost possible for them to bring down the wages of labour to any point at which bare existence can be maintained."[46] If workers were treated this poorly some contemporaries also worried that they could neither work to their full potential nor afford any of the newly manufactured goods. Domestic demand would dry up, to the detriment of business owners. And what was to guarantee that business owners who made huge profits would eventually raise the wages of their workers or reinvest their profits in new businesses and hire more workers in the future? Why would they not spend their profits on themselves? And what would ensure that their profits even would stay in Britain rather than a foreign country? In short, some people questioned whether wealth would trickle down at all. This might have included workers themselves, who could have learned about Smith's ideas from a publication like the *Gorgon*, which described in very clear terms the position of "Adam Smith on the Lower Classes" in *The Wealth of Nations*.[47]

The uncertainty that prosperity would reach the lower classes led some Britons, even businessmen, to look for an alternative to laissez-faire. Robert Owen, for example, who was co-owner of the largest cotton mill in Britain, New Lanark, believed that it was essential for factory owners to provide the best possible conditions for their workers, including good housing and education. This would simultaneously help the workers and the nation as a whole, and it would increase the profits of manufacturers because happy workers were more efficient producers. Eventually, however, Owen concluded that abandoning capitalism and private property for the creation of cooperative communities in which all property was held in common was an even better solution.

Owen's ideas were not shared by the majority of businessmen or the middle or upper classes. More common were ideas like those of Robert Malthus, who argued that the population of Britain was growing so rapidly that it would soon outstrip the food supply. To check this population growth, Malthus argued that Parliament had to change its **Poor Laws** and stop giving financial assistance to the poor.

Under the **Poor Laws** in Britain before 1834, if a worker was not earning enough money to support his family, the local parish would provide him with food and a job, perhaps in a workhouse. In addition, after 1795 under the "Speenhamland system," workers could get handouts of money (known as "outdoor relief") as supplements to top up wages that did not keep up with food prices.

Malthus and many other businessmen argued that government "relief" must stop because it removed the incentive for workers to take any kind of employment they could find, and at whatever wages were offered no matter how low. As a result, it kept alive unuseful members of society, like the sick or aged. The Poor Law also cost an increasing amount of money in taxes known as Poor Rates, paid by landowners and tenants, which were rising with the expanding poverty caused by high food prices and low wages.

But if given no government poor relief, what would become of the industrial workers who could be laid off at the will of their employers and paid as little as possible to work in terrible conditions? The citizen quoted above predicted that eventually "Englishmen will revolt, because they will not lie down quietly and starve by thousands." He concluded that "capital" in the form of industry "will, if not restrained, destroy the whole constitution of England. I believe that it is fighting a war now as much against the palace of the King, and the palaces of the nobles, and the mansions of the manufacturers, as it is against the cottages of the poor; and I believe that the public will not, because they cannot, bear it much longer."[48] But would anyone do anything to "restrain" the new industry?

Clearly, the "engines of mischief" that some workers attacked in the early nineteenth century were not just the new inventions made of iron and wood that propelled the factory system but the entire social, economic, and political system that was threatening their way of life.

CONCLUSION

By 1817 the Industrial Revolution was well underway. But the problems it engendered were very much present. And the people of Britain were divided over how to solve those problems and how to respond to the new industrial economy. Should the new industry be stopped and the British return to their age-old traditions and type of work? Or should new technology and work conditions be expanded and improved? Or was some combination of both possible? These and other fundamental questions had to be considered. The cotton weavers and spinsters, craftsmen, merchants, and aristocrats in 1817 and 1818 Manchester are facing their own crises. How each group responds is the subject of this game.

3 The Game

MAJOR ISSUES FOR DEBATE

In this game students will debate important issues such as the following:

- Industrialization. What was the impact of machine labor and the factory system in Great Britain? How do individuals and society as a whole adapt to radically new technology and the changes it brings? How were the working classes, middle classes, and various propertied interests affected differently? What pressures did technological and economic innovation create for the political, legal, and social systems?
- Labor exploitation. Were the employment models pursued by factory owners exploitative, or a painful step in the dawn of liberty for the hitherto subsistence-level working poor?
- Capitalism and socialism. What are capitalist theories of value and free market exchange? How did early socialists critique the role of the "invisible hand" in distributing social and economic resources? What is the role of the state in regulating the economy?
- Benevolence and self-interest. What is the relationship between benevolence and self-interest? Does a good citizen have the duty to work for the common good of society? What is the common good of society? Is it best to rely on individual action to increase one's wealth and status, or is collective action more effective in achieving those ends? Is benevolence more important than self-interest?
- The nature of class and status. What is the relationship between social status, classes, and money? How is status protected and/or expanded through legal change?
- Political power and powerlessness. How do the powerless seek to appropriate political power? What factors lead to, and inhibit, revolutionary outbreaks? Should citizens rely on peaceful negotiations and political activity, or on violent coercion in order to achieve their ends?

In the game, characters from all social classes will take different positions on these questions.

RULES AND PROCEDURES

Objectives and Victory Conditions

The objective of all players is to successfully negotiate the transition to the new industrial economy. They do this as individuals and as members of society. Successful players will

- help themselves economically and socially as *individuals* and
- advance their class's vision of what is good for *society* as a whole.

In other words, a player does well if they can improve their own material condition and social standing, and at the same time persuade others that what benefits their class contributes to the welfare of the entire British nation. This means not only that successful players will end the game with a higher income and better lifestyle than they began with but also that they will have put into place laws or other mechanisms that contribute to the long-term well-being of their class.

Players carry out these objectives in the two parts of the game: Town Hall and Market Place.

In Town Hall, players must convince the aristocrats to pass laws that benefit their class, themselves, and the English nation by making convincing arguments in speeches and debates, which can form the basis of writings they submit to local newspapers.

Market Place is where players work to improve their individual wealth and social standing. If they can secure more favorable employment, make good business deals, or use the power of their numbers, they can earn enough extra income to move up in the class hierarchy.

Status Points

The success of players as individuals is measured by status points.

Players earn status points for **being persuasive**, for example, by giving excellent speeches in Town Hall or publishing an essay in a newspaper that is considered the best of the session. A player who is elected to Parliament in the election that ends the game also can earn status points.

Status points are also awarded to players who **improve their lifestyle**. If they can afford luxuries beyond necessities, or if they can move up a class level or a class by being able to buy the necessities of that class, they earn status points.

Finally, taking actions to **improve the position of their class and themselves** raises a player's status. If workers join together and do something (for example, a Luddite raid, strike, political meeting) to successfully force their employers or the aristocrats to improve their wages or working conditions, they gain status points. If employers stop those actions, they are the ones to gain status.

But status points can be lost as well as gained. If a player does not make enough income to maintain the necessities of their class, they fall down and lose status. If they act contrary to the interests of their class, or if a group action fails and they get arrested and put in prison they also lose points.

Winning

There can be multiple winners of the game. The "overall" winner is the player who most helps themselves as an individual and at the same time helps others in society. This is the player with the greatest number of status points at the end of the game. "Class winners" are the members of the social class whose cumulative status points are the highest.

OUTLINE OF THE GAME

Location and Time

When the game begins the classroom space becomes the center of downtown Manchester. For centuries, beginning with Manchester's rise as a cloth town in the 1300s, Market Place was the focus of most people's lives. Each week residents of all classes gather in the market to buy and sell goods, swap stories, and listen to the news. As Manchester has grown into a textile powerhouse, numerous churches, inns, pubs, houses, and factories have sprung up nearby, including the following:

Town Hall. This is where the local governmental officials, such as magistrates and tax collectors, work. All official governmental meetings take place at Town Hall.

The Weavers' Arms is a local public house ("pub") that caters to people of the working class. Gentlemen did not enter pubs, preferring to do their deal-making in private homes or clubs.

The Exchange is a new building (1809) where middle-class entrepreneurs in the cotton trade conduct business and socialize. Its News Room is a place to relax and read the local and national newspapers.

The Assembly Rooms on Mosley Street is an elegant building with an imposing ballroom and card-room for high-class social events. The Tea Room has a portrait of the Earl of Derby's father in his parliamentary robes hanging over the fireplace.

Collegiate Church is the large neighborhood church (Church of England). **Oldham Street Chapel** (Methodist) is a small chapel located near the working-class neighborhoods on the north end of the town.

Each session of the game represents one week in 1817–18, but each session is separated by one to four months, depending on the length of the game. For an eight-session game, the time might be as follows:

Session 1: one week in December 1817
Session 2: one week in January 1818
Session 3: one week in February 1818
Session 4: one week in March 1818
Session 5: one week in April 1818
Session 6: one week in May 1818
Session 7: one week in June 1818
Session 8: one week in July 1818

Activities for Each Session

1. Town Hall

The majority of each session takes place in Town Hall, which is presided over by the magistrate. This is where new laws are considered and voted upon. During our game, these laws will apply to all of Britain, not just Manchester. This is because the members of Parliament in London are uncertain about what is best for the new industrial cities, and so have asked our aristocrats to make decisions for them.

England in 1817–18 was not a democracy. At the beginning of the game only male members of the aristocracy can vote, and they are in charge of Town Hall.

But because laws passed in Town Hall are a matter of general interest to all citizens, all players are present, and anyone can speak if called on by the magistrate.

Any citizen can give their opinion about any law. But ultimately their goal is to convince the aristocrats to vote the way they want. This is done with formal and informal speeches, and general debates. Enlisting the aid of others to speech for or against a law may increase the chances of success.

The debates in Town Hall usually focus on whether or not the aristocrats should pass laws to

- establish a minimum wage for handweavers and spinsters, or other form help for them (e.g., a tax on machines);
- regulate the conditions of factories or of child labor;
- reinstate an income tax, and/or change indirect taxes;
- abolish or reform the Corn Laws;
- abolish or reform the Poor Laws; and
- allow more people the right to vote and sit in Parliament.

At the end of Town Hall, the aristocrats can decide to pass laws or not. The magistrate may conduct the voting privately or publicly.

2. Market Place

When Town Hall concludes (perhaps after two-thirds of a session), everyone goes to Market Place. Here they can buy, trade, pay bills, socialize, strategize, and—of course—hear or read the news. While the magistrate presides over Town Hall, no one is in charge of Market Place.

The following should happen each session in Market Place:

- The publican announces what he thinks is the best or most important item (editorial, poem, cartoon, etc.) from a newspaper of the week and reads or summarizes its contents.
- News of important developments is announced by any citizen, such as the opening of a new factory, the invention of a new machine, the foundation of a new charity.
- Vicars can give sermons warning of violations of the Seven Deadly Sins by any citizen and read banns (announcements) of any upcoming marriages.
- Factions have meetings and discussions. The working class generally meets at the Weavers' Arms pub. The middle classes meet in the News Room of the Exchange. And aristocrats retire to the tea rooms of the Assembly Rooms.

Market Place also is where individuals and factions or classes apply the economic ideas of their favorite thinkers to their individual circumstances to improve their income, lifestyle, and status. Different classes do this differently. Each class has specific economic projects to execute, which will give them choices about how to best advance their lifestyle and achieve their long-term interests. Players must use their imagination and also think about the consequences of their actions. For example,

- Aristocrats can
  — rent out the warehouses they own in downtown Manchester to merchants who wish to open a factory;
  — open a mineral (coal, tin, copper, etc.) mine on their estate;
  — invest in a business or give a loan with interest; and
  — open charities to help the poor.
- Merchants can
  — maximize the profits of their existing hand-production business (e.g., by lowering wages, increasing hours, increasing productivity);
  — open a new business (e.g., a factory by renting a warehouse from an aristocrat, buying a machine, and hiring English or immigrant workers—see GM for the latter); and
  — expand an existing business by hiring more workers or buying more machines.
- Craftsmen can
  — make and sell machines; and
  — open a hand or factory business.
- Publicans can
  — sell drink and food in their pubs;
  — find information about the goings on in Manchester; and
  — devise other ways to serve their customers.
- Weavers and spinsters can strategize about how to get higher wages, lower hours, improved working and living conditions (e.g., by helping to get laws passed, negotiating with employers, organizing a strike or work slowdown, or trying to get the right to vote).
- Vicars can
  — look for new parishioners;
  — help existing parishioners make more income or raise their status; and
  — open a workhouse, school, factory.
- Editors can
  — solicit items to publish in the next edition of their newspapers; and
  — strategize with their political supporters to help them achieve their goals.

Anything else players may conceive is possible, as long as it is historically accurate. Players should ask the GM for permission and provide evidence that it could have occurred in Manchester in 1817–18.

Accounting

Manchester is a big city even in 1817, so it will be necessary that both the players and the game master keep track of all the public and private changes that occur over the course of the game. Some developments, like laws passed in Town Hall, can impact entire classes of people. Other changes, such as individual deals made in Market Place, only effect a few.

Players will be responsible for keeping track of those things that impact them as individuals. The game master will give players worksheets on which they indicate how much money they made that session and how much they are required to spend on necessities to be part of a certain social class. If players have extra income, they can save it to use in future sessions, or they can spend it on luxuries, which may help them rise up a class or level in the social hierarchy and earn status points. If they cannot afford necessities, they will fall down into a lower social class or level and lose status points.

Here are some of the things players must know:

- Their **weekly income** each session. This can change if wages are raised or lowered, a player buys or sells goods (raw cotton, finished textiles, machines), or players make or take business investments or loans. The instructor will provide a worksheet to help players calculate how much money they make each session.
- The **lifestyle expenses** they will pay each session. Players must decide how much of their income to spend on necessities and how much they can use for luxuries. This also determines if they can gain status points for living a better lifestyle than others at their level, or if they can move up a level or class. Again, the instructor will provide a worksheet to help players make these calculations.
- The **status points** they earned that session. These status points can come from a player's improving their lifestyle, making excellent speeches in Town Hall, getting the best article that session published in a newspaper, or taking group actions.

The game master will keep track of the following that impact all citizens of Manchester:

- **Market Conditions Board**. This lists the current wages, rents, prices (of finished textiles, machines, and cotton), interest rates, operating expenses, and the productivity of various businesses. These are necessary for players to calculate their weekly income. They may change each session based on laws passed in Town Hall, or in response to market forces (for example, a bad harvest, international competition, oversaturation of the market) at the discretion of the GM.
- **Relationships.** This includes who is working for whom in various types of businesses.
- **Laws or Actions**. Any laws passed in Town Hall or actions taken by individual players (for example, opening a factory).
- **Leaderboard.** This is a running total of the status points and personal income of all players.

Possible Other Events

Other things may happen in the course of a session. For example, there might be a collective action of weavers or spinsters such as a machine-breaking raid or political meeting. A trial for any of these actions or for breaking the Combination Law also could occur. Anything that was possible historically is possible in the game.

General Election

The game culminates with the general election of 1818, which took place throughout the spring and summer of that year. This election is to choose the two members representing the county of Lancashire (where Manchester is located) in the House of

Commons of the national Parliament in London. These MPs will be able vote on all laws for all of England, making them positions of very real importance and status.

Who can vote for, or run in, the election of members of Parliament:

- All male members of the aristocracy can vote in these elections and all, except the Earl of Derby, can run for a seat as a member of Parliament.
- Other players can vote in the election if they are able to buy a certain amount of land (costing 800s). They can run in the election for MP if they own even more land (costing 8,000s).

Therefore, to gain the right to vote or run in this election, players must have quite a bit of extra money and convince an aristocrat to sell them the land necessary—the aristocracy owns almost all the land in England. Because land was still the source of great wealth, power, and status in 1818, aristocrats should be reluctant to sell. Those who do sell land will lose status points.

Candidates for election must give speeches outlining their positions on key issues and making arguments for why they should be elected. Those players who cannot vote are still able to influence the elections by asking difficult questions, heckling, making counterspeeches, or taking direct action, all of which was very common at the time.

The winners of the election earn extra status points.

EXAMPLE SCHEDULE OF SESSIONS

The schedule below is for eleven class sessions of seventy-five minutes each (four setup days, six game days, one debriefing day). The instructor may modify it to include more or less days, and for longer or shorter class times.

Setup 1: Introduce eighteenth-century England: social classes and political system, Industrial Revolution: causes, nature, consequences.
Setup 2: Ideas of Adam Smith, Malthus.
Setup 3: Ideas of Owen, Poor Laws, problems of handweavers, Luddites. Describe game mechanics.
Setup 4: Demands for political reform. Explain Market Conditions Board and accounting. Introduce players and hold first faction meetings. Quiz (optional).
Game Session 1: Debate on handweavers' condition and what to do about it, such as a minimum wage or tax on machines.
Game Session 2: Debates on factory conditions and child labor.
Game Session 3: Debates on taxation and Poor Laws.
Game Session 4: Debates on political reforms.
Game Session 5: Open debates/electioneering.
Game Session 6: Election!
Debriefing: What really happened, and lessons learned.

THE MANY FUNCTIONS OF THE GAME MASTER

Sometimes, the game master functions as a cheerleader, encouraging factions and individuals to work hard to win the game. Sometimes the GM functions as an instructor, explaining to students how they can give more effective speeches and research and write better papers. The GM may also perform other functions, such as posting and adjusting the market conditions. Other functions of the GM include

- announcing events impacting Manchester, such as a factory accident, epidemic diseases, a poor harvest, the arrival of more immigrants, lower prices for textiles because of foreign competition;
- determining the consequences of player actions and unforeseen events with die rolls; and
- playing the role of a character outside Manchester, such as Lord Byron or the prime minister, or even the prince regent.

THE GAME MASTER NEWS SERVICE

The game master will provide some clues as to "where" the class is in real history by issuing Game Master News Service updates periodically. For example, the Game Master News Service might report about a food riot or Luddite trial in the neighborhood. A by-election for Parliament or election results in other counties may make the news. All of these can serve as topics for debate in Town Hall or Market Place, as well as indicate the date of the game.

ASSIGNMENTS AND GRADING

This game requires that students advance their objectives by writing papers, giving speeches, participating in debates, and conducting negotiations.

At minimum students should expect to give one formal speech and write one formal essay. Depending on the length of the game and requirements of the instructor, more speeches and essays may be required.

The instructor will announce the total number of pages that each student must write for the game, and also indicate the relative importance of each type of assignment (for example, written work may comprise two-thirds of the grade, and class participation one-third).

Writing assignments can consist of articles, editorials, cartoons, advertisements, and so on that will appear in the two newspapers, which will be published several times during the game (the actual number of issues will depend on the length of the game).

Players are expected to speak frequently during the game. Sometimes they will speak extemporaneously in response to debates or negotiations, and at other times they will give speeches from the podium. Reading speeches aloud without delivering a speech effectively is discouraged; those who do so will likely receive grade penalties or reductions in status points.

Students must support their written and spoken views with more facts and other forms of evidence than are contained in their role sheets or this game book. This means that for most papers and speeches, students must conduct research to strengthen their work. The game book contains a selected bibliography; many other issues can be researched online. The Industrial Revolution is one of the great subjects of historical inquiry.

4
Roles and Factions

Roles in this game include representatives of all major social classes. Regency England was very class-conscious place. Everyone was keenly aware of what class they were born into and what their place was in society. This is not to say that a person could not transcend their birth class; in fact, class stratification and class transcendence was a compelling notion during this time and, as such, was one of the major themes of fiction and literature from the period.

SOCIAL CLASSES

Society was divided into a hierarchy with a few people at the top controlling most of the wealth and power, the majority at the bottom with very little, and some in the middle.

For the purposes of the game, characters are placed into this class structure based on how much income they earn each week. With this income they must purchase necessities befitting their social class. In other words, they must be able to afford the necessities of a class to be considered part of it. If they earn any extra income, they can gain status by being able to afford luxuries. Even more money may allow them to pay for the necessities of a higher level in their class or even an entire social class, and they can move up in the hierarchy.

The goal of players in the game is to improve their lifestyle by affording luxuries and advancing up the social ladder. They also must ensure that laws are passed in Town Hall that will help them maintain their status in the future.

There are three classes and three levels within each class in the game based on a character's weekly income:

1. **Working classes**
   - working poor
   - respectable working class
   - labor aristocracy
2. **Middle classes**
   - lower middle class
   - middle middle class
   - upper middle class

3. **Upper classes**
   - squires
   - patricians
   - magnates

FACTIONS

In addition to social classes, there are three initial factions in the game based on general agreement about ideas. These factions usually work together in Town Hall debates. Players may change factions or create new factions as the game progresses.

1. **Aristocratic faction**, including
   - members of the aristocracy
   - conservative editor
   - Church of England vicar
2. **Middle-class faction**, including
   - merchants
   - craftsmen
3. **Working-class faction**, including
   - weavers and spinsters
   - liberal editor
   - Methodist vicar

INDETERMINACY

Unlike some Reacting to the Past games, no characters in *Engines of Mischief* are either fully indeterminate or fully determinate. In other words, all characters have strong opinions on certain subjects (are determinate about them) but are undecided (indeterminate) on other topics and thus are open to skillful persuasion. As a result, all members of a faction may not be in entire agreement on certain things. This means that individual players are encouraged to think of ways to influence their faction, as well as members of other factions, to achieve their goals. But they should not forget that a balance must be reached between helping themselves, their social class, and Britain as a whole.

Secrets

Even though all characters are part of a faction, which works to advance itself and its members, it is important to note that some roles have secrets. Players with secrets should be very careful not to reveal them to anyone unless they can use them to advance their status. And everyone should be cautious to guard against a secret being used against them.

ROLES

The character roles in the game and their place in the social classes of Regency England are as follows:

Aristocracy

In England the highest social class was the aristocracy. It was divided into two groups: peers and gentry. A peer was the eldest male in a family who had legal title to nobility. The gentry included all others in the family of peers and also landed gentlemen with lesser titles. The two groups were closely related, but the gentry did not have any legal rights different from other Britons.

Aristocrats made a living primarily from the land they inherited. While they sold some produce grown on their land and sometimes sold minerals mined on the land, most of their income came from rents collected from their tenants who were farmers. The ideal for an aristocrat was not to work but to oversee the management of his estate and perhaps some investments, and to participate in governing the country. However, unlike the aristocracy in continental Europe, there was no such thing as "derogation"; British aristocrats did not lose their titles if they got involved in business, and they were always on the lookout for new sources of wealth. This made the English aristocracy the richest in the world.

It also gave them a great deal of power and nearly complete control over both national and local government. Aristocrats thought of themselves as near the top of the "Great Chain of Being" that descended from God, though the various ranks of humans, down to the lowest insect. They expected their inferiors to accept their superiority in the social hierarchy without question, to keep to their own lowly places, and display due deference.

The Peers

The peers owned by far the most land of any class other than royalty. For the most part, the titles that the peers held were passed down through inheritance (as in hereditary peerages) or given for life (as in life peerages). Following the tradition of primogeniture, the eldest son inherited both the legal noble title and all of the land. This is why the English nobility was the richest in the world.

Hereditary titles in order of their rank are as follows: duke (duchess), marquess (marchioness), earl (countess), viscount (viscountess), baron or baronet (baroness). There also were life peerages that made a person noble for life but that were not inherited by their family upon death.

In England, the peers were granted the right to sit in the House of Lords and given power by the sovereign (king or queen) to select justices of the peace (also known as magistrates) locally. However, many peers eschewed local matters, preferring to spend as time in London much as possible. Their huge country estates were left in the care of bailiffs or land agents, who took on the responsibility of renting the land to their tenant farmers.

In this game, we have a peer and the son of a peer. Our peer has been named the lord lieutenant of the county of Lancashire (like a US state) in which the borough (town) of Manchester is located. He is responsible for enforcing the will of the king. The first duty of the lord lieutenant is to work with the magistrate (another player) to keep the peace.

Earl Edward Smith-Stanley, Twelfth Earl of Derby. The Earl of Derby is an aristocrat and the wealthiest and most important person in all of Manchester. He is legally a peer, and socially a magnate. Because of this, he was chosen by the king to be the lord lieutenant of the county. He owns a townhouse in London where he spends most of his time, but he also has an extensive landed estate in Manchester that he often visits. He likes to throw parties and spend a great deal of money on himself, his family, and his friends. He is a member of the House of Lords and thus cannot run for election to the House of Commons in Parliament.

The Gentry

Just below peers in social standing are the gentry, sometimes called *landed gentry* to reflect the fact that they owned land. These were individuals from the top tier of the upper class. They also rented land to tenant farmers and managed their estates similarly to the peers above them.

The gentry were educated, comfortably well-off gentlemen who spent much of their time trying to elevate their social standing. They tended to stay nearer to their estates and traveled less frequently than the wealthier landowners. As such, they were active in local affairs, and often were chosen by the local peer to be a magistrate (justice of the peace) or to hold other local offices.

Since there was no local police force, the magistrates were expected to enforce laws, hold trials, and keep the peace. To assist in these tasks, magistrates could call in the local militia or national army to help quell disturbances or to enforce the law. The magistrates also had the ability to read the Riot Act when crowds got unruly. The Riot Act was first passed in 1714 by Parliament to give local magistrates a way to control crowds. A magistrate could read the Riot Act to any group of twelve or more people assembled in one place, at which time the group had to disperse peacefully within twenty minutes or be held accountable for rioting, a capital crime. (A capital crime means that the perpetrator is subject to death if found guilty.) (The full text of the Riot Act is given in the core texts in this gamebook.)

In our game, the peers and gentry are the only players eligible to sit for the position of magistrate or vote in Town Hall meetings. At the beginning of the game, they also are the only players who can vote in, or run for, parliamentary elections because they are the only players who meet the qualifications of landownership.

William Lamb, Second Viscount Melbourne. Viscount Melbourne is a single, thirty-five-year-old member of the gentry. Socially he is of the patrician class. He has a comfortable, genteel life on a vast manor estate. His hobbies include falconry, riding, fox hunting, butterfly collecting, and plant propagation. He is extremely well-read and well-educated and interested in government. He especially wants to help the earl govern Manchester as well as possible. He is able to run for election to the House of Commons in Parliament.

Lady Shirley Keeldar. Lady Keeldar is twenty-five years old, single, and very rich. She also is of the patrician social class. She grew up in India, where her aristocratic father served in the army. She had to return to England after attending finishing school in Switzerland, because both her parents died tragically in India. She inherited a large estate near Manchester that she manages all on her own. She very much values her independence and wants to maintain the wealth that ensures it. She is not able to vote in Town Hall or run for election to the House of Commons in Parliament because she is a woman.

Baronet Sir Thomas Joseph Trafford. As the third son of a peer, Baronet Trafford had little hope of inheriting his family's large estate, so at the age of sixteen he joined the Royal Navy. After being rewarded by the king with a pension for his service in the Napoleonic Wars and in the War of 1812, he purchased a manor house and many acres of land near Manchester. This makes him a gentleman in social class rank as well as a member of the gentry. He is deeply loyal to the Crown to whom he owes so much. Because he is so trustworthy, Lord Derby named him the first major-commandant of the Manchester and Salford Yeomanry cavalry, a militia formed in 1817 because of fear of growing working-class unrest. He also is very interested in finding ways to make more money. He is able to run for election to the House of Commons in Parliament.

Baronet Sir Oswald Mosley. Oswald Mosley is the second Baronet Mosley and the apple of his mother's eye. He also is a member of the gentry and part of the gentleman social class. Rather than stay home and manage the estate, as his father (the first Baronet) wanted him to do, he decided to attend Oxford University. There he met Percy Bysshe Shelley and his circle of Romantic poets. He traveled to Germany and Switzerland with some of them and dabbles in poetry himself. Now that he is back in Manchester, he has inherited his father's house and estate, which he has not managed particularly well. He lives comfortably but is desperate to find ways to finance his own poetry career and the art of all of his friends. He is able to run for election to the House of Commons in Parliament.

Most aristocrats in the game are determinate (convinced) about the need for their class to maintain its control over the political system. They think the established tradition of government by the aristocracy is best. They accept the ideas of paternalism and "noblesse oblige," in other words that noble people have an obligation to take care of their inferiors as fathers care for their children. They also prefer the older economic system of agricultural production on the land that they mostly own. They are indeterminate (less certain) about the new industry and technology. British aristocrats were always on the lookout for new sources of income. Although owning or operating a factory was beneath them socially, they might be open to investment opportunities and helping those middle-class businessmen who advance the wealth of the nation.

Middle Classes

As in society as a whole, there was a hierarchy among the middle classes. Middle-class men did many different jobs. They acted as bankers, as merchants trading goods over long distances, and as manufacturers, professionals, and shopkeepers. Some members of the middle class could be very wealthy, while others were just getting by. What was common to all, however, was that they worked with their brains, rather than their hands, and they had to have a good deal of education and training to perform their duties. This is why merchants, craftsmen, vicars, and editors are considered part of the middle classes.

Merchants and manufacturers represent the entrepreneurship required to fuel a capitalist economy. While they were hardly gentlemen or members of the gentry, they aspired to be so. As these men became more and more successful in generating capital, they often desired landownership rights, and many wished to spend their growing sums of money to live like the aristocracy. But land was not always easy to come by. Nevertheless, a struggling member of the peerage or landed gentry might see fit to settle debts through the sale or transfer of some of their landed property.

Where craftsmen fit in the social hierarchy was not always clear. While they could come from poor working families and did not need to attend a formal school to do their jobs, they did have training and were considered skilled workers. Craftsmen in trades with apprenticeships worked their ways up from apprentice to journeyman, to small master, and finally master craftsman. Apprenticeships were one way that a young man of the poorer, laboring classes could move into the higher-status lower middle class.

For the purposes of this game the craftsmen are part of the middle-class faction. This is because in the early Industrial Revolution, many of the merchants, the mill and factory owners who were considered solidly middle class, grew out of the craftsman class. Usually they were of "small master" rank, and they coupled their artisanal skills with a certain measure of risk-taking and entrepreneurship by opening their own shops and even creating new inventions to improve and expand their businesses. In the course of the game, therefore, our craftsmen eventually are able to do the same things as our merchants.

The vicars and editors in the game are members of the middle class because of their social backgrounds and education. However, their political views, and the clients they serve, place them in either the aristocratic or working-class factions and they share those factions' determinacy or indeterminacy.

Those characters in the middle-class faction in the game are determinate about advancing themselves economically; that is their primary concern. However, they are indeterminate about how exactly to do this. They all like the new technology, but they are not sure whether to put all their eggs into one basket, the factory system, or whether to continue with hand production, either exclusively or in addition to opening a factory. And they are not entirely sure how best to maximize their profit. Should they ruthlessly follow the ideas of Adam Smith and keep wages and costs low, or should they adopt some of the theories of Robert Owen and be kind to their human machines?

Merchants

In our game, we have several merchants. They all are of the middle middle class in terms of the social hierarchy, but they can have slightly different jobs. They are manufacturers, bankers, and traders. They can work together as a faction in Town Hall as well as compete with one another in Market Place.

Samuel Oldknow. At the age of sixty-one Samuel Oldknow is one of the elder statesmen of textile production. He was born into a comfortable middle-class family that had run a successful drapery business and was apprenticed as a teenager to learn the trade. By the age of twenty-five he was able to use his family connections to establish himself as a cotton manufacturer in Manchester. He was one of the first to try using new hand machines in a large workshop and even a steam engine to power spinning mules. This business failed, and he returned to hand production on the putting-out system. But he is very interested in trying to use machines in a factory again.

Hannah Thornton. Hannah Thornton is very unusual; she is one of the very few women running a cotton textile business in England. Her husband started the Thornton Company before they married, and when he died a few years ago she took over running it by herself. She is determined to make it a success, so that she can turn it over to her teenage son (when he comes of age) in a condition that will make him immensely wealthy. She is devoted to her son and is ruthless in looking out for his interests.

John Kennedy. John Kennedy runs a cotton textile business. He had to work his way up from almost nothing, having grown up on a small farm in Scotland. His father apprenticed him when he was fourteen to a nearby cotton manufacturer, where he learned all the details of the trade and about the new machines that were just starting to be used.

After years in Manchester, Kennedy was able to open his own spinning mill powered by a steam engine. Unfortunately, he went bankrupt just six years ago when a downturn in trade coincided with the Luddite machine-breaking raids. He has started over in the hand production of textiles in cottages. He is not entirely sure whether to risk using machines in a factory setting again.

Hugh Hornby Birley. Hugh Hornby Birley comes from a well-off merchant family that has been making textiles for years. He has three brothers and four sisters and is determined to do better than all of them combined. With money given him by his wealthy father, he started a textile business and will do whatever he can to make it succeed. He also has strong political opinions, being a staunch Tory (Conservative). When Sir Thomas Trafford was appointed commander of the Manchester Yeomanry, Birley immediately volunteered to be a captain, and he quite likes parading around in uniform on his horse with a sword.

Benjamin Heywood. Benjamin Heywood is a banker. He is partner in Benjamin Heywood, Sons & Co. Bank, also known as Manchester Bank, which was established by his grandfather and is now run by his uncle. His is only twenty-five years old and is very concerned to show his uncle that he has the makings of an excellent banker and should become the director of the bank instead of his younger brothers. He is not afraid to advance loans for any scheme that will make his bank a profit.

William Rathbone V. William Rathbone is an importer of raw cotton. He divides his time between the port city of Liverpool, where raw cotton is brought to Britain, and Manchester, where he sells that cotton to textile merchants. He is always looking for cheaper sources of raw cotton, and ways to get manufacturers to produce more and more textiles. He is well aware that cotton is produced in the United States by slave labor and would prefer a different source of cotton but is not sure where to find it or if he can afford it. He is interested in the possibilities the new industry has for expanding his business.

James M'Connel. James M'Connel also is a merchant who arranges for the production and sale of cotton textiles. His father was a simple Scottish farmer, but M'Connel had much higher ambitions. As a merchant he has moved into the middle class, and he wants to stay there. He is very interested in the promise that the new machines hold for his ability to make more money and is not opposed to using immigrant and child labor. But he worries about the real possibility of failure that he has seen happen to others who took such risks, so he might consider sticking with handloom weaving.

Craftsmen

The craftsmen in our game make different products that require skill, training, and ingenuity, from wheels, to clocks, to wooden goods, to metal objects. The manager of our local pub also is a craftsman. They are in the lower middle class socially, but most of them share many ideas with others in the middle classes and so are part of that faction in Town Hall.

Richard Roberts. Richard Roberts is a wheelwright. His father, who was a shoemaker, had him educated at a Sunday school and secured him an apprenticeship so that he could make perfect wheels. But Roberts was always interested in the wheels used to spin cotton. He made a beautiful one with inlaid woods for his mother. Now he is interested in using his wheel-making skills to invent a self-acting mule that will improve the process of cotton spinning.

William Horrocks. William Horrocks is a clockmaker. His father was a handweaver who wanted his son to have a better future, so he apprenticed him to a master clockmaker, where he learned to work with gears, glass, metal, and wood. But weaving remained very important to him. He is particularly interested in power-looms and hopes to invent a compact one made out of metal that cannot be destroyed by Luddites.

Peter Blinstone. Peter Blinstone is a blacksmith who forges metal items from iron and steel. He uses hand tools such as large hammers, tongs, chisels, and anvils. He was apprenticed at a young age and progressed through journeyman status to finally being the master of his own blacksmithing shop. He is physically strong and accustomed to working hard.

Neil Snodgrass. Neil Snodgrass is a carpenter, originally from a poor farming family in Scotland. He learned how to construct things out of wood, especially houses, carts, and rough furniture after years of being an apprentice. Upon his move to Manchester, he became interested in textile production. Now he is using his carpenter skill to work on an invention—a scutching or blowing machine, for opening and cleaning raw cotton or flax.

John Reid. John Reid is the proprietor of the Weavers' Arms, a local public house ("pub"), which his father ran before him, and his father before him. He serves working-class men pints of ale and exotic cocktails, provides pretty barmaids for atmosphere, and gives prostitutes a place to ply their wares. He began allowing some militant prolabor revolutionaries to meet in the pub as a way to make some easy money, and now his pub is the center of radical activity for the entire borough.

Vicars

Also in the middle classes are the vicars, or parish priests, of Manchester. But because vicars do not make very large incomes, they are part of the lower middle class socially. There are two of them with slightly different religious and political views, which is why one is part of the aristocratic faction and the other part of the working-class faction in Town Hall.

The official church in England is the Anglican church, or simply the Church of England. The Church of England was founded in 1534 during the reign of Henry VIII after the Roman Catholic pope refused the king's request for a divorce from his wife, Catherine of Aragon. Henry VIII decided to renounce the pope and found his own church. This church is "Protestant" because it was founded as a protest against the pope and also against many of his "Catholic" doctrines and practices. The British monarch is the official head of the Church of England.

In 1817, the local unit of the Church of England was the *parish*. In cities, like Manchester, the parish was roughly equivalent to a district or a neighborhood. The clergyman in charge of the parish building and the church members was called the *rector*. The rector sometimes appointed a *vicar* to govern in his stead. The rector or vicar maintained the parish church buildings where he held religious services. He usually lived in a parsonage (vicar's home) next to it. Vicars often held Sunday schools and other educational programming, and sometimes provided poor relief.

In the first half of the 1700s, a young Anglican priest named John Wesley began an evangelical spiritual movement within the Church of England that came to be known as Methodism. Founded in response to a perception of apathy in the Church of England, the hallmarks of the Methodist movement were that lay (nonordained) members of the church should be able to preach and teach, and that every person has equal worth in the eyes of God—regardless of his or her social standing. As such, a primary goal of the early Methodist church was to help the downtrodden, the needy, and the sick. Wesley was also an abolitionist and a supporter of prison reform. Many Methodist preachers were derided by traditional priests as fanatics for their loud singing and enthusiastic preaching.

By the early nineteenth century, some followers of Methodism (though certainly not all) took the advice to form societies led by lay preachers further than Wesley had probably intended. They were able to self-organize at the grassroots level to create agencies for feeding the poor and for evangelizing to the unsaved. The resulting egalitarianism and "can do" spirit found in some Methodist groups could be quite frightening to the established, traditional Church of England. This egalitarianism was especially appealing to the working classes and especially scary to the aristocracy. As historian E. P. Thompson explains, one prevailing view was that "if Christ's poor came to believe that their souls were as good as aristocratic or bourgeois souls then it might lead them on to the arguments of [Thomas Paine's] *Rights of Man*."[1] Paine's book of course presented numerous arguments against hereditary rule and in favor of both the French and American Revolutions. Upon its publication (1791–92) it was immediately banned in England.

In our game we have two vicars, one from each of these types of Anglicanism: a traditionalist and a Methodist.

Charles Wickstead Ethelston. Charles Ethelston is a vicar in the Church of England in charge of a large Anglican church in Manchester. He is a firm believer in tradition and hierarchy. He fears the new belief in equality will disrupt a social structure that has provided for order and stability in England from time immemorial. He thinks the poor must accept God's plan for them and do their duties with thankfulness and piety. He is a strong supporter of the aristocracy and their right to govern Manchester in whatever way they see fit, and thus begins the game in their faction.

William Bramwell. William Bramwell is a radical minister who became a Methodist. He believes that all people are sinners equally, so rich and poor are not fundamentally different. Because of this Bramwell sympathizes with the working classes and their struggle against the oppressive rule of the traditional church and the entrenched aristocracy. He is very much opposed to slavery in the British colonies. He oversees a small Methodist chapel in the heart of Manchester. Because of his political sympathies Vicar Bramwell begins the game in the working-class faction.

Editors

The game is set in a time period when the press was extremely important. Increasing numbers of people could read, and those who could not had papers read to them in pubs and coffeehouses. Newspapers were popular and influential. In the early nineteenth century there were fifty-two newspapers in London and over 100 newspapers elsewhere in England. Some newspapers printed just the news and advertisements. Others did not try to be objective but openly represented the views of the different political parties. There are two newspaper editors in our game. Like the vicars, although socially they both are of the lower middle class, in Town Hall they are part of different factions.

Charles Wheeler. Charles Wheeler is the founder of a traditional pro-king, progovernment newspaper called the *Manchester Chronicle*. This newspaper is one of the four Tory papers in Manchester. Even though it is considered dull, it is the most popular Conservative paper in Manchester, with over 3,000 copies sold a week. Wheeler is stodgy, virulently anti-Luddite, antiradical, and unfailingly progovernment. His motto is "church and king," and he doesn't stand for any radical activity. He thinks that the people are ignorant and foolish, and that they need his help to get to the right way of thinking. He most definitely is part of the aristocratic faction in the game.

James Wroe. James Wroe is the publisher of the *Manchester Observer*. It is a labor-friendly and working-class-friendly weekly newspaper. Wroe originally came from the lower classes; his father was a pub owner. He eventually grew to believe that the state was passing laws, such as taxes on newspapers and state-mandated minimum prices for newspapers, which were designed to keep the working man from having any power so that the entrenched aristocracy could rule England undisturbed. That is when he decided to found his own radical newspaper specifically for the workers. It is a popular newspaper, selling over 4,000 copies per week. He is part of the working-class faction in the game.

Working Classes

The members of the working classes in our game are urban dwellers; they live in the city of Manchester. But there is a pecking order, or a status hierarchy, within the working class. There were distinctions between men who engaged in the trades (shopkeepers), men who performed skilled labor (such as carpenters, bricklayers, tanners, butchers, millwrights, and weavers), and the lowest classes of unskilled labor such as those doing farm labor, mining, or menial tasks. Higher-status occupations commanded a higher labor price. They sometimes were even known as the "labor aristocracy."

The relative status of working-class occupations at this time can be understood according to three factors: the nature and amount of education or training required to perform that job; whether the worker is self-employed or receives a wage; and how much physical labor is involved in the job.

The weavers and spinsters in our game consider

themselves part of a higher class of laborers. Although they work with their hands, their ancestors had been members of guilds who were trained through apprenticeships and had independence because they worked at home and controlled their working conditions and the product of their labor.

However, with the recent decline in demand for handweaving textiles and the competition of factories and machines, our weavers and spinsters are now receiving a meager wage from a merchant employer and giving the products they make to their "boss" to do with as he or she sees fit. So, they no longer are working for themselves or controlling their work product. But they still have the independence of working at home.

Because of their straightened circumstances, the weavers and spinsters in our game are part of the respectable working classes, rather than the labor aristocracy where they think they should belong. But if their wages decline further, they may fall down into the working poor and be forced to take less desirable work. Jobs that were paid on a case-by-case or contract basis were notoriously underpaid and unreliable, and easily could result in destitution in bad times. People with dangerous, dirty, or uncomfortable jobs such as colliers (coal miners), farm laborers, ratcatchers, chimney sweeps, and rag and bone sellers, definitely were lower in status than regular factory workers. In many cases, immigrant labor gravitated first to these low-paying, low-skilled jobs. Our weavers and spinsters are very much afraid of being reduced to this level.

So the working classes in our game are definitely determined to improve their work and home lives. However, they are indeterminate on the tactics to use to do this. They can choose to work in the new factories, where they will earn a decent and regular wage but will lose even more independence and may have to put up with dangerous and unhealthy working conditions. Or they can try to force their employers or the government to raise their wages and maintain good work conditions by uniting together and taking collective action. If they do this, they must decide what type of action to take. Should they smash the machines as Luddites, form a secret union and plan a strike, or concentrate on securing the right to participate in the political system?

In our game, we have both men and women who start out producing handmade cotton textiles in their homes.

Weavers
- Robert Ellison
- Samuel Bamford
- John Bromilow
- Thomas Kerfoot
- William Scholfield
- William Baines
- Robert Pilkington
- Elijah Dixon
- William Finn
- Joseph Bruce

These men are all weavers by training, just as their fathers and grandfathers were. They work mostly in cotton, but occasionally in wool or silk. They do their weaving on a loom in their attic at home, working with the women and children of the family, who card the raw wool and then spin it into the thread for the men to weave. They sell the finished product to a merchant who comes by the house each week with more raw cotton. The weavers are not entirely illiterate, many can read and write, but they also love to hear the newspapers read at the pub. As a result, they are well aware of developments in the world around Manchester.

Spinsters
- Mary Cannon
- Lydia Molyneux
- Susannah Sexton
- Betty Armstrong
- Ethelinda Wilson
- Ann Scott
- Martha Partington
- Alice McGrath
- Ellen Croft

FIGURE 4.1 Joseph Cooper, *Weavers' Cottages*, 1971. © Bury Art Museum, Greater Manchester, UK.

These women perform an important function in the hand production of cloth. They spin raw cotton in the attics of their homes on a spinning wheel or a spinning jenny. They work with the children of the family, who do the carding of raw cotton with hand-held tools, and the men of the family, who take the thread they spin and weave it into cloth on a loom. The spinsters do the cooking, cleaning, and childcare at home, as well as the spinning. However, they also get the chance to go to the Weavers' Arms to hear the newspapers read.

Note: In our game, our weavers and spinsters all are unmarried, but they do not live alone. They reside in part of a "cottage" that is a three-story townhouse in the center of Manchester similar to the one pictured in figure 4.1.

In their cottage are one or three other family members with whom they work—fathers, mothers, sisters, brothers, aunts, uncles, or cousins. Players in the game "control" the labor of the other residents in their cottage. Initially, they all work for the same employer. But if our weaver or spinster decides to change employers or type of employment (such as to join a factory) or participate in a collective action (Luddite raid, strike, political meeting), the other members of the cottage will join them.

Thus, although each player is a single worker, they represent the interests of far more members of their class. This is to reflect the fact that in historical reality there were far more workers than people in the middle and upper classes.

5
Core Texts

Adam Smith, *The Wealth of Nations*

Adam Smith wrote The Wealth of Nations *in order to answer a question dear to the hearts of monarchs throughout Europe: What factors and economic policies make nations wealthier? In the process of answering this question, Smith developed a robust account of how markets work, their advantages, and some of their pitfalls. He wrote at the dawn of the Industrial Revolution, so although his theories provided a framework in which merchants conceived of their contributions to British society, he would not have been personally aware of the impact of large-scale factory operations on society. In our readings, we will see Smith develop several crucially important ideas:*

- *The division of labor—the role of specialization and invention in increasing the productivity of human economic efforts.*
- *Selfishness and beneficial social outcomes ("the invisible hand")—the claim that self-interested human exchanges (such as buying goods in the market) play a far greater role in increasing social welfare than directly benevolent undertakings (such as giving money to someone in need).*
- *Natural and market prices of goods. Smith gives an account of the natural prices of goods in terms of their average costs plus anticipated profit, and then discusses how supply and demand for goods in markets can exceed or fall short of their natural price.*
- *The role of government in correcting for negative market effects. Smith is aware of potential distortions created by markets, or by the laws that undergird them. Pay particular attention to his account of the uneven bargaining positions of workers and factory owners, as well as his account of the role the state should play in public education.*

Here are some questions for your character consider as you read through this document:

- *Why does Smith think that self-interested market exchanges tend to increase social well-being? What, in your character's experience, are examples that will challenge or support this claim?*
- *How does thinking about the price of labor in terms of market prices support or undermine the dignity of workers? Is it humane to treat human labor as a commodity? Is it possible to treat it otherwise?*
- *What role does Smith think the state should play in correcting for negative effects of markets? Look especially at his argument for how the division of labor can harm the well-being of workers who spend their entire day doing simple and repetitive tasks. Why does Smith think the state should counteract this mentally deadening effects with public education? What implications does this argument have for other questions, such as the conditions of factories, the status of unions, and/or the minimum wage?*

Source: Adam Smith, *An Inquiry into the Nature and Causes of the Wealth of Nations* (London: W. Strahan and T. Cadell, 1776).

BOOK 1, Chapter 1. Of the Division of Labour

... THE greatest improvement in the productive powers of labour, and the greater part of the skill, dexterity, and judgment with which it is anywhere directed, or applied, seem to have been the effects of the division of labour.

[The famous example of the pin factory.]

To take an example, ... from a very trifling manufacture; but one in which the division of labour has been very often taken notice of, the trade of the pinmaker; a workman not educated to this business (which the division of labour has rendered a distinct trade), nor acquainted with the use of the machinery employed in it (to the invention of which the same division of labour has probably given occasion), could scarce, perhaps, with his utmost industry, make one pin in a day, and certainly could not make twenty. But in the way in which this business is now carried on, not only the whole work is a peculiar trade, but it is divided into a number of branches, of which the greater part are likewise peculiar trades.

One man draws out the wire, another straights it, a third cuts it, a fourth points it, a fifth grinds it at the top for receiving, the head; to make the head requires two or three distinct operations; to put it on is a peculiar business, to whiten the pins is another; it is even a trade by itself to put them into the paper; and the important business of making a pin is, in this manner, divided into about eighteen distinct operations, which, in some manufactories, are all performed by distinct hands, though in others the same man will sometimes perform two or three of them.

[How much more effective this system is than without the division of labor.]

I have seen a small manufactory of this kind where ten men only were employed, and where some of them consequently performed two or three distinct operations. But though they were very poor, and therefore but indifferently accommodated with the necessary machinery, they could, when they exerted themselves, make among them about twelve pounds of pins in a day. There are in a pound upwards of four thousand pins of a middling size. Those ten persons, therefore, could make among them upwards of forty-eight thousand pins in a day. Each person, therefore, making a tenth part of forty-eight thousand pins, might be considered as making four thousand eight hundred pins in a day. But if they had all wrought separately and independently, and without any of them having been educated to this peculiar business, they certainly could not each of them have made twenty, perhaps not one pin in a day; that is, certainly, not the two hundred and fortieth, perhaps not the four thousand eight hundredth part of what they are at present capable of performing, in consequence of a proper division and combination of their different operations.

[The division of labor can be applied to other manufacturing industries but is less possible in agriculture, where it is difficult to give different jobs to different people.]

[Why does the division of labor work?]

... This great increase of the quantity of work

which, in consequence of the division of labour, the same number of people are capable of performing, is owing to three different circumstances; first, to the increase of dexterity in every particular workman; secondly, to the saving of the time which is commonly lost in passing from one species of work to another; and lastly, to the invention of a great number of machines which facilitate and abridge labour, and enable one man to do the work of many. . . .
[Smith expands upon how the division of labor results in improvements in the dexterity and efficiency of workers, and in technological improvements to machines made by those workers.]

[With the division of labor, everyone can make more than they need. This opens the possibility of markets for trading surplus.]
. . . It is the great multiplication of the productions of all the different arts, in consequence of the division of labour, which occasions, in a well-governed society, that universal opulence which extends itself to the lowest ranks of the people. Every workman has a great quantity of his own work to dispose of beyond what he himself has occasion for; and every other workman being exactly in the same situation, he is enabled to exchange a great quantity of his own goods for a great quantity, or, what comes to the same thing, for the price of a great quantity of theirs. He supplies them abundantly with what they have occasion for, and they accommodate him as amply with what he has occasion for, and a general plenty diffuses itself through all the different ranks of the society. . . .

BOOK 1, Chapter 2. Of the Principle Which Gives Occasion to the Division of Labour

THIS division of labour, from which so many advantages are derived, is not originally the effect of any human wisdom, which foresees and intends that general opulence to which it gives occasion. It is the necessary, though very slow and gradual consequence of a certain propensity in human nature which has in view no such extensive utility; the propensity to truck, barter, and exchange one thing for another.

[Animals use persuasion to get what they want.]
. . . It is common to all men, and to be found in no other race of animals, which seem to know neither this nor any other species of contracts. . . . Nobody ever saw a dog make a fair and deliberate exchange of one bone for another with another dog. Nobody ever saw one animal by its gestures and natural cries signify to another, this is mine, that yours; I am willing to give this for that. When an animal wants to obtain something either of a man or of another animal, it has no other means of persuasion but to gain the favour of those whose service it requires. A puppy fawns upon its dam, and a spaniel endeavours by a thousand attractions to engage the attention of its master who is at dinner, when it wants to be fed by him.

[People also do this.]
Man sometimes uses the same arts with his brethren, and when he has no other means of engaging them to act according to his inclinations, endeavours by every servile and fawning attention to obtain their good will. He has not time, however, to do this upon every occasion. In civilised society he stands at all times in need of the cooperation and assistance of great multitudes, while his whole life is scarce sufficient to gain the friendship of a few persons. In almost every other race of animals each individual, when it is grown up to maturity, is entirely independent, and in its natural state has occasion for the assistance of no other living creature.

[But man will be more successful when he gets what he wants AND the giver also gets something in return. The introduction to Smith's concept of "self-interest." This also includes the very famous passage about "the butcher, the brewer, or the baker."]
But man has almost constant occasion for the help of his brethren, and it is in vain for him to expect it from their benevolence only. He will be more likely to prevail if he can interest their self-love in his favour, and show them that it is for their own advantage to do for him what he requires of them. Whoever offers to another a bargain of any kind, proposes to do this. Give me that which I want, and you shall have this which you want, is the meaning of every

such offer; and it is in this manner that we obtain from one another the far greater part of those good offices which we stand in need of. It is not from the benevolence of the butcher, the brewer, or the baker that we expect our dinner, but from their regard to their own interest. We address ourselves, not to their humanity but to their self-love, and never talk to them of our own necessities but of their advantages. Nobody but a beggar chooses to depend chiefly upon the benevolence of his fellow-citizens. . . .

[How self-interest leads to better business.]

. . . In a tribe of hunters or shepherds a particular person makes bows and arrows, for example, with more readiness and dexterity than any other. He frequently exchanges them for cattle or for venison with his companions; and he finds at last that he can in this manner get more cattle and venison than if he himself went to the field to catch them. From a regard to his own interest, therefore, the making of bows and arrows grows to be his chief business, and he becomes a sort of armourer. Another excels in making the frames and covers of their little huts or movable houses. He is accustomed to be of use in this way to his neighbours, who reward him in the same manner with cattle and with venison, till at last he finds it his interest to dedicate himself entirely to this employment, and to become a sort of house-carpenter. In the same manner a third becomes a smith or a brazier, a fourth a tanner or dresser of hides or skins, the principal part of the nothing of savages. And thus the certainty of being able to exchange all that surplus part of the produce of his own labour, which is over and above his own consumption, for such parts of the produce of other men's labour as he may have occasion for, encourages every man to apply himself to a particular occupation, and to cultivate and bring to perfection whatever talent or genius he may possess for that particular species of business. . . .

BOOK 1, Chapter 5. Of the Real and Nominal Price of Commodities, or Their Price in Labour, and Their Price in Money

[The division of labor leads to a market price for labor.]

EVERY man is rich or poor according to the degree in which he can afford to enjoy the necessaries, conveniences, and amusements of human life. But after the division of labour has once thoroughly taken place, it is but a very small part of these with which a man's own labour can supply him. The far greater part of them he must derive from the labour of other people, and he must be rich or poor according to the quantity of that labour which he can command, or which he can afford to purchase. The value of any commodity, therefore, to the person who possesses it, and who means not to use or consume it himself, but to exchange it for other commodities, is equal to the quantity of labour which it enables him to purchase or command. Labour, therefore, is the real measure of the exchangeable value of all commodities.

The real price of everything, what everything really costs to the man who wants to acquire it, is the toil and trouble of acquiring it. What everything is really worth to the man who has acquired it, and who wants to dispose of it or exchange it for something else, is the toil and trouble which it can save to himself, and which it can impose upon other people. . . . Labour was the first price, the original purchase-money that was paid for all things. It was not by gold or by silver, but by labour, that all the wealth of the world was originally purchased; and its value, to those who possess it, and who want to exchange it for some new productions, is precisely equal to the quantity of labour which it can enable them to purchase or command. . . .

BOOK 1, Chapter 7. Of the Natural and Market Price of Commodities

[Ordinary rates.]

THERE is in every society or neighbourhood an ordinary or average rate both of wages and profit in every different employment of labour and stock.

This rate is naturally regulated, as I shall show hereafter, partly by the general circumstances of the society, their riches or poverty, their advancing, stationary, or declining condition; and partly by the particular nature of each employment.

There is likewise in every society or neighbourhood an ordinary or average rate of rent, which is regulated too, as I shall show hereafter, partly by the general circumstances of the society or neighbourhood in which the land is situated, and partly by the natural or improved fertility of the land.

[Ordinary, natural, average.]

These ordinary or average rates may be called the natural rates of wages, profit, and rent, at the time and place in which they commonly prevail.

[Natural prices.]

When the price of any commodity is neither more nor less than what is sufficient to pay the rent of the land, the wages of the labour, and the profits of the stock employed in raising, preparing, and bringing it to market, according to their natural rates, the commodity is then sold for what may be called its natural price.

[Profit is necessary.]

The commodity is then sold precisely for what it is worth, or for what it really costs the person who brings it to market; for though in common language what is called the prime cost of any commodity does not comprehend the profit of the person who is to sell it again, yet if he sell it at a price which does not allow him the ordinary rate of profit in his neighbourhood, he is evidently a loser by the trade; since by employing his stock in some other way he might have made that profit. His profit, besides, is his revenue, the proper fund of his subsistence. As, while he is preparing and bringing the goods to market, he advances to his workmen their wages, or their subsistence; so he advances to himself, in the same manner, his own subsistence, which is generally suitable to the profit which he may reasonably expect from the sale of his goods. Unless they yield him this profit, therefore, they do not repay him what they may very properly be said to have really cost him. . . .

[Market price.]

The actual price at which any commodity is commonly sold is called its market price. It may either be above, or below, or exactly the same with its natural price.

[Effectual demand.]

The market price of every particular commodity is regulated by the proportion between the quantity which is actually brought to market, and the demand of those who are willing to pay the natural price of the commodity, or the whole value of the rent, labour, and profit, which must be paid in order to bring it thither. . . .

[Supply and demand, competition, and price gouging.]

When the quantity of any commodity which is brought to market falls short of the effectual demand, all those who are willing to pay the whole value of the rent, wages, and profit, which must be paid in order to bring it thither, cannot be supplied with the quantity which they want. Rather than want it altogether, some of them will be willing to give more. A competition will immediately begin among them, and the market price will rise more or less above the natural price, according as either the greatness of the deficiency, or the wealth and wanton luxury of the competitors, happen to animate more or less the eagerness of the competition. Among competitors of equal wealth and luxury the same deficiency will generally occasion a more or less eager competition, according as the acquisition of the commodity happens to be of more or less importance to them. Hence the exorbitant price of the necessaries of life during the blockade of a town or in a famine. . . .

[Markets keep prices "natural."]

The natural price, therefore, is, as it were, the central price, to which the prices of all commodities are continually gravitating. Different accidents may sometimes keep them suspended a good deal above it, and sometimes force them down even somewhat below it. But whatever may be the obstacles which hinder them from settling in this centre of repose and continuance, they are constantly tending towards it. . . .

[Using apprenticeship laws to keep skilled wages high.]

The same statutes of apprenticeship and other corporation laws indeed, which, when a manufacture is in prosperity, enable the workman to raise his wages a good deal above their natural rate, sometimes oblige him, when it decays, to let them down a good deal below it. As in the one case they exclude many people from his employment, so in the other they exclude him from many employments. The effect of such regulations, however, is not near so durable in sinking the workman's wages below, as in raising them above their natural rate. Their operation in the one way may endure for many centuries, but in the other it can last no longer than the lives of some of the workmen who were bred to the business in the time of its prosperity. When they are gone, the number of those who are afterwards educated to the trade will naturally suit itself to the effectual demand. The police must be as violent as that of Indostan or ancient Egypt (where every man was bound by a principle of religion to follow the occupation of his father, and was supposed to commit the most horrid sacrilege if he changed it for another), which can in any particular employment, and for several generations together, sink either the wages of labour or the profits of stock below their natural rate.

BOOK 1, Chapter 8. Of the Wages of Labour

THE produce of labour constitutes the natural recompense or wages of labour.

[What is the proper wage for this labor.]

What are the common wages of labour, depends everywhere upon the contract usually made between those two parties, whose interests are by no means the same. The workmen desire to get as much, the masters to give as little as possible. The former are disposed to combine in order to raise, the latter in order to lower the wages of labour.

[Why masters have an advantage over workmen.]

It is not, however, difficult to foresee which of the two parties must, upon all ordinary occasions, have the advantage in the dispute, and force the other into a compliance with their terms. The masters, being fewer in number, can combine much more easily; and the law, besides, authorizes, or at least does not prohibit their combinations, while it prohibits those of the workmen. We have no acts of parliament against combining to lower the price of work; but many against combining to raise it. In all such disputes the masters can hold out much longer. A landlord, a farmer, a master manufacturer, a merchant, though they did not employ a single workman, could generally live a year or two upon the stocks which they have already acquired. Many workmen could not subsist a week, few could subsist a month, and scarce any a year without employment. In the long run the workman may be as necessary to his master as his master is to him; but the necessity is not so immediate.

[How masters combine without appearing to do so.]

We rarely hear, it has been said, of the combinations of masters, though frequently of those of workmen. But whoever imagines, upon this account, that masters rarely combine, is as ignorant of the world as of the subject. Masters are always and everywhere in a sort of tacit, but constant and uniform combination, not to raise the wages of labour above their actual rate. To violate this combination is everywhere a most unpopular action, and a sort of reproach to a master among his neighbours and equals. We seldom, indeed, hear of this combination, because it is the usual, and one may say, the natural state of things, which nobody ever hears of. Masters, too, sometimes enter into particular combinations to sink the wages of labour even below this rate. These are always conducted with the utmost silence and secrecy, till the moment of execution, and when the workmen yield, as they sometimes do, without resistance, though severely felt by them, they are never heard of by other people.

[Workers combine to retaliate.]

Such combinations, however, are frequently resisted by a contrary defensive combination of the workmen; who sometimes too, without any provocation of this kind, combine of their own accord to raise the price of their labour. Their usual pretences are, sometimes the high price of provisions; sometimes the great profit which their masters make by their work. But whether their combinations be offensive or defensive, they are always abundantly heard of. In order to bring the point to a speedy decision, they have always recourse to the loudest clamour, and sometimes to the most shocking violence and outrage. They are desperate, and act with the folly and extravagance of desperate men, who must either starve, or frighten their masters into an immediate compliance with their demands. The masters upon these occasions are just as clamorous upon the other side, and never cease to call aloud for the assistance of the civil magistrate, and the rigorous execution of those laws which have been enacted with so much severity against the combinations of servants, labourers, and journeymen. The workmen, accordingly, very seldom derive any advantage from the violence of those tumultuous combinations, which, partly from the interposition of the civil magistrate, partly from the necessity superior steadiness of the masters, partly from the necessity which the greater part of the workmen are under of submitting for the sake of present subsistence, generally end in nothing, but the punishment or ruin of the ringleaders.

[How low can you go? The masters will try to go as low as possible.]

But though in disputes with their workmen, masters must generally have the advantage, there is, however, a certain rate below which it seems impossible to reduce, for any considerable time, the ordinary wages even of the lowest species of labour.

A man must always live by his work, and his wages must at least be sufficient to maintain him. They must even upon most occasions be somewhat more; otherwise it would be impossible for him to bring up a family, and the race of such workmen could not last beyond the first generation. . . .

The liberal reward of labour, therefore, as it is the necessary effect, so it is the natural symptom of increasing national wealth. The scanty maintenance of the labouring poor, on the other hand, is the natural symptom that things are at a stand, and their starving condition that they are going fast backwards. . . .

[Whether it is good to pay working people enough money.]

Is this improvement in the circumstances of the lower ranks of the people to be regarded as an advantage or as an inconveniency to the society? The answer seems at first sight abundantly plain. Servants, labourers, and workmen of different kinds, make up the far greater part of every great political society. But what improves the circumstances of the greater part can never be regarded as an inconveniency to the whole. No society can surely be flourishing and happy, of which the far greater part of the members are poor and miserable. It is but equity, besides, that they who feed, clothe, and lodge the whole body of the people, should have such a share of the produce of their own labour as to be themselves tolerably well fed, clothed, and lodged.

[Poverty, marriage, and sex in the lower classes.]

Poverty, though it no doubt discourages, does not always prevent marriage. It seems even to be favourable to generation. A half-starved Highland woman frequently bears more than twenty children, while a pampered fine lady is often incapable of bearing any, and is generally exhausted by two or three. Barrenness, so frequent among women of fashion, is very rare among those of inferior station. Luxury in the fair sex, while it inflames perhaps the passion for enjoyment, seems always to weaken, and frequently to destroy altogether, the powers of generation.

[Poverty and child mortality.]

But poverty, though it does not prevent the generation, is extremely unfavourable to the rearing of children. The tender plant is produced, but in so cold a soil and so severe a climate, soon withers and dies. . . . Though their marriages are generally more fruitful than those of people of fashion, a smaller proportion of their children arrive at maturity. In foundling hospitals, and among the children brought up by parish charities, the mortality is still greater than among those of the common people.

Every species of animals naturally multiplies in proportion to the means of their subsistence, and no species can ever multiply beyond it. But in civilised society it is only among the inferior ranks of people that the scantiness of subsistence can set limits to the further multiplication of the human species; and it can do so in no other way than by destroying a great part of the children which their fruitful marriages produce. . . .

[High wages keep populations high.]

The liberal reward of labour, therefore, as it is the effect of increasing wealth, so it is the cause of increasing population. To complain of it is to lament over the necessary effect and cause of the greatest public prosperity.

[Economic growth in a society is best for the working class.]

It deserves to be remarked, perhaps, that it is in the progressive state, while the society is advancing to the further acquisition, rather than when it has acquired its full complement of riches, that the condition of the labouring poor, of the great body of the people, seems to be the happiest and the most comfortable. It is hard in the stationary, and miserable in the declining state. The progressive state is in reality the cheerful and the hearty state to all the different orders of the society. The stationary is dull; the declining, melancholy.

[High wages also encourage more productivity. But too much productivity is a bad thing. Rest days are required. Vacations are good and natural.]

The liberal reward of labour, as it encourages the propagation, so it increases the industry of the common people. The wages of labour are the encouragement of industry, which, like every other human quality, improves in proportion to the encouragement it receives. A plentiful subsistence increases the bodily strength of the labourer, and the comfortable hope of bettering his condition, and of ending his days perhaps in ease and plenty, animates him to exert that strength to the utmost. Where wages are high, accordingly, we shall always find the workmen more active, diligent, and expeditious than where they are low. . . . Some workmen, indeed, when they can earn in four days what will maintain them through the week, will be idle the other three. This, however, is by no means the case with the greater part. Workmen, on the contrary, when they are liberally paid by the piece, are very apt to overwork themselves, and to ruin their health and constitution in a few years. . . . It is the call of nature, which requires to be relieved by some indulgence, sometimes of ease only, but sometimes, too, of dissipation and diversion. If it is not complied with, the consequences are often dangerous, and sometimes fatal, and such as almost always, sooner or later, brings on the peculiar infirmity of the trade. If masters would always listen to the dictates of reason and humanity, they have frequently occasion rather to moderate than to animate the application of many of their workmen. It will be found, I believe, in every sort of trade, that the man who works so moderately as to be able to work constantly not only preserves his health the longest, but, in the course of the year, executes the greatest quantity of work. . . .

[In good times, entrepreneurship flourishes and wages rise.]

In years of plenty, servants frequently leave their masters, and trust their subsistence to what they can make by their own industry. But the same cheapness of provisions, by increasing the fund which is destined for the maintenance of servants, encourages masters, farmers especially, to employ a greater number. Farmers upon such occasions expect more profit from their corn by maintaining a few more labouring servants than by selling it at a low price in the market. The demand for servants increases, while the number of those who offer to supply that demand diminishes. The price of labour, therefore, frequently rises in cheap years.

[Everyone wants a wage in tough times. Entrepreneurship suffers. Wages sink.]

In years of scarcity, the difficulty and uncertainty of subsistence make all such people eager to return to service. But the high price of provisions, by diminishing the funds destined for the maintenance of servants, disposes masters rather to diminish than to increase the number of those they have. In dear years, too, poor independent workmen frequently consume the little stocks with which they had used to supply themselves with the materials of their work, and are obliged to become journeymen for subsistence. More people want employment than can easily get it; many are willing to take it upon lower terms than ordinary, and the wages of both servants and journeymen frequently sink in dear years....

[How employers play against each other in times of plenty.]

In a year of sudden and extraordinary plenty, there are funds in the hands of many of the employers of industry sufficient to maintain and employ a greater number of industrious people than had been employed the year before; and this extraordinary number cannot always be had. Those masters, therefore, who want more workmen bid against one another, in order to get them, which sometimes raises both the real and the money price of their labour.

[How workers play against each other in times of scarcity.]

The contrary of this happens in a year of sudden and extraordinary scarcity. The funds destined for employing industry are less than they had been the year before. A considerable number of people are thrown out of employment, who bid against one another, in order to get it, which sometimes lowers both the real and the money price of labour. In 1740, a year of extraordinary scarcity, many people were willing to work for bare subsistence. In the succeeding years of plenty, it was more difficult to get labourers and servants....

BOOK 1, Chapter 10. Of Wages and Profit in the Different Employments of Labour and Stock

[How people choose different jobs.]

... THE whole of the advantages and disadvantages of the different employments of labour and stock must, in the same neighbourhood, be either perfectly equal or continually tending to equality. If in the same neighbourhood, there was any employment evidently either more or less advantageous than the rest, so many people would crowd into it in the one case, and so many would desert it in the other, that its advantages would soon return to the level of other employments....

PART I. INEQUALITIES ARISING FROM THE NATURE OF THE EMPLOYMENTS THEMSELVES

[Why certain jobs can command higher wages than others.]

... THE five following are the principal circumstances which, so far as I have been able to observe, make up for a small pecuniary gain in some employments, and counterbalance a great one in others: first, the agreeableness or disagreeableness of the employments themselves; secondly, the easiness and cheapness, or the difficulty and expense of learning them; thirdly, the constancy or inconstancy of employment in them; fourthly, the small or great trust which must be reposed in those who exercise them; and, fifthly, the probability or improbability of success in them.

[(1) Wages vary with their ease, cleanliness, etc.]

First, the wages of labour vary with the ease or hardship, the cleanliness or dirtiness, the honourableness or dishonourableness of the employment. Thus in most places, take the year round, a journeyman tailor earns less than a journeyman weaver. His work is much easier. A journeyman weaver earns less than a journeyman smith. His work is not always easier, but it is much cleanlier. A journeyman blacksmith, though an artificer, seldom earns so much in twelve hours as a collier, who is only a labourer, does in eight. His work is not quite so dirty, is less dangerous, and is carried on in daylight, and above ground. Honour makes a great part of the reward of all honourable professions. In point of pecuniary gain, all things considered, they are generally under-recompensed, as I shall endeavour to show by and by. Disgrace has the contrary effect. The trade of a butcher is a brutal and an odious business; but it is in most places more profitable than the greater part of common trades. The most detestable of all employments, that of public executioner, is, in proportion to the quantity of work done, better paid than any common trade whatever. *[And in advanced societies, people will do for pleasure what was once done for a living.]*

[(2) Wages vary with how easy the business is to learn. Apprenticeships keep wages high.]

Secondly, the wages of labour vary with the easiness and cheapness, or the difficulty and expense of learning the business....

The difference between the wages of skilled labour and those of common labour is founded upon this principle.

The policy of Europe considers the labour of all mechanics, artificers, and manufacturers, as skilled labour; and that of all country labourers as common labour. It seems to suppose that of the former to be of a more nice and delicate nature than that of the latter.... The laws and customs of Europe, therefore, in order to qualify any person for exercising the one species of labour, impose the necessity of an apprenticeship, though with different degrees of rigour in different places. They leave the other free and open to everybody. During the continuance of the apprenticeship, the whole labour of the apprentice belongs to his master. In the meantime he must, in many cases, be maintained by his parents or relations, and in almost all cases must be clothed by them. Some money, too, is commonly given to the master for teaching him his trade. They who cannot give money give time, or become bound for more than the usual number of years; a consideration which, though it is not always advantageous to the master, on account of the usual idleness of apprentices, is always disadvantageous to the apprentice....

[Jobs that require more study should be paid more.]

Education in the ingenious arts and in the liberal professions is still more tedious and expensive. The pecuniary recompense, therefore, of painters and sculptors, of lawyers and physicians, ought to be much more liberal; and it is so accordingly....

[(3) Steady employment will drive wages down but will be made up for in regular pay.]

Thirdly, the wages of labour in different occupations vary with the constancy or inconstancy of employment. Employment is much more constant in some trades than in others. In the greater part of manufacturers, a journeyman may be pretty sure of employment almost every day in the year that he is able to work. A mason or bricklayer, on the contrary, can work neither in hard frost nor in foul weather, and his employment at all other times depends upon the occasional calls of his customers. He is liable, in consequence, to be frequently without any. What he earns, therefore, while he is employed, must not only maintain him while he is idle, but make him some compensation for those anxious and desponding moments which the thought of so precarious a situation must sometimes occasion. Where the computed earnings of the greater part of manufacturers, accordingly, are nearly upon a level with the day wages of common labourers, those of masons and bricklayers are generally from one half more to double those wages.... The high wages of those workmen, therefore, are not so much the recompense of their skill, as the compensation for the inconstancy of their employment....

[(4) Does the job require a lot of trust? Goldsmiths, physicians, lawyers.]

Fourthly, the wages of labour vary accordingly to the small or great trust which must be reposed in the workmen. The wages of goldsmiths and jewellers are everywhere superior to those of many other workmen, not only of equal, but of much superior ingenuity, on account of the precious materials with which they are intrusted.

We trust our health to the physician: our fortune and sometimes our life and reputation to the lawyer and attorney. Such confidence could not safely be reposed in people of a very mean or low condition. Their reward must be such, therefore, as may give them that rank in the society which so important a trust requires. The long time and the great expense which must be laid out in their education, when combined with this circumstance, necessarily enhance still further the price of their labour....

[(5) Wages rise if it is likely that a great deal of skill and intelligence is required.]

Fifthly, the wages of labour in different employments vary according to the probability or improbability of success in them.

The probability that any particular person shall ever be qualified for the employment to which he is educated is very different in different occupations. In the greater part of mechanic trades, success is almost certain; but very uncertain in the liberal professions. Put your son apprentice to a shoemaker, there is little doubt of his learning to make a pair of shoes; but send him to study the law, it is at least twenty to one if ever he makes such proficiency as will enable him to live by the business. . . .

PART II. INEQUALITIES BY THE POLICY OF EUROPE

[Having discussed the differences in the nature of jobs themselves, Smith moves on to discuss how government policies make it worse by establishing apprenticeship laws, etc.]

. . . THE policy of Europe, by not leaving things at perfect liberty, occasions other inequalities of much greater importance.

It does this chiefly in the three following ways. First, by restraining the competition in some employments to a smaller number than would otherwise be disposed to enter into them; secondly, by increasing it in others beyond what it naturally would be; and, thirdly, by obstructing the free circulation of labour and stock, both from employment to employment and from place to place.

[The government requires apprenticeships.]

First, the policy of Europe occasions a very important inequality in the whole of the advantages and disadvantages of the different employments of labour and stock, by restraining the competition in some employments to a smaller number than might otherwise be disposed to enter into them.

The exclusive privilege of an incorporated trade necessarily restrains the competition, in the town where it is established, to those who are free of the trade. To have served an apprenticeship in the town, under a master properly qualified, is commonly the necessary requisite for obtaining this freedom. The [bylaws] of the corporation regulate sometimes the number of apprentices which any master is allowed to have, and almost always the number of years which each apprentice is obliged to serve. The intention of both regulations is to restrain the competition to a much smaller number than might otherwise be disposed to enter into the trade. The limitation of the number of apprentices restrains it directly. A long term of apprenticeship restrains it more indirectly, but as effectually, by increasing the expense of education. . . . Seven years seem anciently to have been, all over Europe, the usual term established for the duration of apprenticeships in the greater part of incorporated trades.

[The government is encouraging employment in some fields but not others.]

Secondly, the policy of Europe, by increasing the competition in some employments beyond what it naturally would be, occasions another inequality of an opposite kind in the whole of the advantages and disadvantages of the different employments of labour and stock. . . . It has been considered as of so much importance that a proper number of young people should be educated for certain professions, that sometimes the public and sometimes the piety of private founders have established many pensions, scholarships, exhibitions, bursaries, etc., for this purpose, which draw many more people into those trades than could otherwise pretend to follow them. *[Smith gives the example of churchmen, lawyers, and physicians.]*

[The government restricts people from switching their line of work or relocating out of their parish if they become poor.]

Thirdly, the policy of Europe, by obstructing the free circulation of labour and stock both from employment to employment, and from place to place, occasions in some cases a very inconvenient inequality in the whole of the advantages and disadvantages of their different employments.

The Statute of Apprenticeship obstructs the free circulation of labour from one employment to another, even in the same place. The exclusive privileges of corporations obstruct it from one place to another, even in the same employment. . . .

The obstruction which is . . . given to [the free circulation of labor] by the Poor Laws is, so far as I know, peculiar to England. It consists in the difficulty which a poor man finds in obtaining a settlement, or even in being allowed to exercise his industry in any parish but that to which he belongs. It is the labour of artificers and manufacturers only of which the free circulation is obstructed by corporation laws. The difficulty of obtaining settlements obstructs even that of common labour. It may be worth while to give some account of the rise, progress, and present state of this disorder, the greatest perhaps of any in the police of England.

BOOK 4, Chapter 2. Of Restraints upon the Importation "from Foreign Countries of Such Goods" as Can Be Produced at Home

[Government regulation of foreign trade gives monopolies to certain domestic industries. While this helps those specific industries, it does not benefit the entire nation.]

BY restraining, either by high duties, or by absolute prohibitions, the importation of such goods from foreign countries as can be produced at home, the monopoly of the home-market is more or less secured to the domestick industry employed in producing them. Thus the prohibition of importing either live cattle or salt provisions from foreign countries secures to the graziers of Great Britain the monopoly of the home-market for butchers-meat. The high duties upon the importation of corn, which in times of moderate plenty amount to a prohibition, give a like advantage to the growers of that commodity. The prohibition of the importation of foreign woollens is equally favourable to the woollen manufacturers. . . .

That this monopoly of the home-market frequently gives great encouragement to that particular species of industry which enjoys it, and frequently turns towards that employment a greater share of both the labour and stock of the society than would otherwise have gone to it, cannot be doubted. But whether it tends either to increase the general industry of the society, or to give it the most advantageous direction, is not, perhaps, altogether so evident. . . .

[It is better for the government not to interfere in trade and let businessmen decide how best to make a profit.]

No regulation of commerce can increase the quantity of industry in any society beyond what its capital can maintain. It can only divert a part of it into a direction into which it might not otherwise have gone; and it is by no means certain that this artificial direction is likely to be more advantageous to the society than that into which it would have gone of its own accord.

Every individual is continually exerting himself to find out the most advantageous employment for whatever capital he can command. It is his own advantage, indeed, and not that of the society, which he has in view. But the study of his own advantage naturally, or rather necessarily leads him to prefer that employment which is most advantageous to the society.

First, every individual endeavours to employ his capital as near home as he can, and consequently as much as he can in the support of domestick industry; provided always that he can thereby obtain the ordinary, or not a great deal less than the ordinary profits of stock.

Thus upon equal or nearly equal profits, every wholesale merchant naturally prefers the home-trade to the foreign trade of consumption, and the foreign trade of consumption to the carrying trade. In the home-trade his capital is never so long out of his sight as it frequently is in the foreign trade of consumption. He can know better the character and situation of the persons whom he trusts, and if he should happen to be deceived, he knows better the laws of the country from which he must seek redress. In the carrying trade, the capital of the merchant is, as it were, divided between two foreign countries, and no part of it is ever necessarily brought home, or placed under his own immediate view and command. . . .

A merchant, in the same manner, who is engaged in the foreign trade of consumption, when he collects goods for foreign markets, will always be glad, upon equal or nearly equal profits, to sell as great a part of them at home as he can. He saves himself the risk and trouble of exportation, when, so far as he can, he thus converts his foreign trade of consumption into a home-trade. Home is in this manner the center, if I may say so, round which the capitals of the inhabitants of every country are continually circulating, and towards which they are always tending, though by particular causes they may sometimes be driven off and repelled from it towards more distant employments. But a capital employed in the home-trade, it has already been shown, necessarily puts into motion a greater quantity of domestic industry, and gives revenue and employment to a greater number of the inhabitants of the country, than an equal capital employed in the foreign trade of consumption: and one employed in the foreign trade of consumption has the same advantage over an equal capital employed in the carrying trade. Upon equal, or only nearly equal profits, therefore, every individual naturally inclines to employ his capital in the manner in which it is likely to afford the greatest support to domestick industry, and to give revenue and employment to the greatest number of people of his own country.

Secondly, every individual who employs his capital in the support of domestick industry, necessarily endeavours so to direct that industry, that its produce may be of the greatest possible value....

But it is only for the sake of profit that any man employs a capital in the support of industry; and he will always, therefore, endeavour to employ it in the support of that industry of which the produce is likely to be of the greatest value, or to exchange for the greatest quantity either of money or of other goods....

As every individual, therefore, endeavours as much as he can both to employ his capital in the support of domestick industry, and so to direct that industry that its produce may be of the greatest value; every individual necessarily labours to render the annual revenue of the society as great as he can. He generally, indeed, neither intends to promote the publick interest, nor knows how much he is promoting it. By preferring the support of domestick to that of foreign industry, he intends only his own security; and by directing that industry in such a manner as its produce may be of the greatest value, he intends only his own gain, and he is in this, as in many other cases, led by an invisible hand to promote an end which was no part of his intention. Nor is it always the worse for the society that it was no part of it. By pursuing his own interest he frequently promotes that of the society more effectually than when he really intends to promote it. I have never known much good done by those who affected to trade for the publick good. It is an affectation, indeed, not very common among merchants, and very few words need be employed in dissuading them from it.

What is the species of domestick industry which his capital can employ, and of which the produce is likely to be of the greatest value, every individual, it is evident, can, in his local situation, judge much better than any statesman or lawgiver can do for him. The statesman, who should attempt to direct private people in what manner they ought to employ their capitals, would not only load himself with a most unnecessary attention, but assume an authority which could safely be trusted, not only to no single person, but to no council or senate whatever, and which would nowhere be so dangerous as in the hands of a man who had folly and presumption enough to fancy himself fit to exercise it.

To give the monopoly of the home-market to the produce of domestick industry, in any particular art or manufacture, is in some measure to direct private people in what manner they ought to employ their capitals, and must, in almost all cases, be either a useless or a hurtful regulation. If the produce of domestick can be brought there as cheap as that of foreign industry, the regulation is evidently useless. If it cannot, it must generally be hurtful. It is the maxim of every prudent master of a family, never to attempt to make at home what it will cost him more to make than to buy. The taylor does not attempt to make his own shoes, but buys them of the shoemaker. The shoemaker does not attempt to make his own cloaths, but employs a taylor. The farmer attempts to make neither the one nor the other, but employs those different artificers. All of them find it for their interest to employ their whole industry in a way in which they have some advantage over their neighbours, and to purchase with a part of its produce, or what is the same thing, with the price of a part of it, whatever else they have occasion for?

What is prudence in the conduct of every private family, can scarce be folly in that of a great kingdom. If a foreign country can supply us with a commodity cheaper than we ourselves can make it, better buy it of them with some part of the produce of our own industry, employed in a way in which we have some advantage.

[The government does not need to help businesspeople to act in a self-interested way, because they usually already know to do this, and they will pursue their self-interest best if there are few government regulations to trade. This freedom of trade will contribute to the wealth of the nation as a whole.]

The interest of a nation in its commercial relations to foreign nations is, like that of a merchant with regard to the different people with whom he deals, to buy as cheap and to sell as dear as possible. But it will be most likely to buy cheap, when by the most perfect freedom of trade it encourages all nations to bring to it the goods which it has occasion to purchase; and, for the same reason, it will be most likely to sell dear, when its markets are thus filled with the greatest number of buyers. The act of navigation, it is true, lays no burden upon foreign ships that come to export the produce of British industry. . . . But if foreigners, either by prohibitions or high duties, are hindered from coming to sell, they cannot always afford to come to buy; because coming without a cargo, they must lose the freight from their own country to Great Britain. By diminishing the number of sellers, therefore, we necessarily diminish that of buyers, and are thus likely not only to buy foreign goods dearer, but to sell our own cheaper, than if there was a more perfect freedom of trade.

BOOK 5, Chapter 1. Of the Expenses of the Sovereign or Commonwealth

[Although the government should not play a large role in directing the economy, it does have other important duties that require a certain amount of expense, although how much expense differs in various historical periods.]

THE first duty of the sovereign, that of protecting the society from the violence and invasion of other independent societies, can be performed only by means of a military force. But the expense both of preparing this military force in time of peace, and of employing it in time of war, is very different in the different states of society, in the different periods of improvement. . . .

The second duty of the sovereign, that of protecting, as far as possible, every member of the society from the injustice or oppression *of* every other member of it, or the duty of establishing an exact administration of justice requires too very different degrees of expense in the different periods of society. . . .

The third and last duty of the sovereign or commonwealth is that of erecting and maintaining those public institutions and those public works, which, though they may be in the highest degree advantageous to a great society, are, however, of such a nature, that the profit could never repay the expense to any individual or small number of individuals, and which it therefore cannot be expected that any individual or small number of individuals should erect or maintain. The performance of this duty requires too very different degrees of expense in the different periods of society. . . .

[The third duty of the sovereign includes public works (such as large building projects like roads and bridges), and the education of those who cannot afford private education. Smith here explains why the government should spend money on the education of workers.]

In the progress of the division of labour, the employment of the far greater part of those who live by labour, that is, of the great body of the people, comes to be confined to a few very simple operations; frequently to one or two. But the understandings of the greater part of men are necessarily formed by their ordinary employments. The man whose whole life is spent in performing a few simple operations, of which the effects too are, perhaps, always the same, or very nearly the same, has no occasion to exert his understanding, or to exercise his invention in finding out expedients for removing difficulties which never occur. He naturally loses, therefore, the habit of such exertion, and generally becomes as stupid and ignorant as it is possible for a human creature to become. The torpor of his mind renders him, not only incapable of relishing or bearing a part in any rational conversation, but of conceiving any generous, noble, or tender sentiment, and consequently of forming any just judgment concerning many even of the ordinary duties of private life. Of the great and extensive interests of his country he is altogether incapable of judging; and unless very particular pains have been taken to render him otherwise, he is equally incapable of defending his country in war. The uniformity of his stationary life naturally corrupts the courage of his mind, and makes him regard with abhorrence the irregular, uncertain, and adventurous life of a soldier. It corrupts even the activity of his body, and renders him incapable of exerting his strength with vigour and perseverance, in any other employment than that to which he has been bred. His dexterity at his own particular trade seems, in this manner, to be acquired at the expense of his intellectual, social, and martial virtues.

But in every improved and civilized society this is the state into which the labouring poor, that is, the great body of the people, must necessarily fall, unless government takes some pains to prevent it.

It is otherwise in the barbarous societies, as they are commonly called, of hunters, of shepherds, and even of husbandmen in that rude state of husbandry which precedes the improvement of manufactures, and the extension of foreign commerce. In such societies the varied occupations of every man oblige every man to exert his capacity, and to invent expedients for removing difficulties which are continually occurring. Invention is kept alive, and the mind is not suffered to fall into that drowsy stupidity, which, in a civilized society, seems to benumb the understanding of almost all the inferior ranks of people. In those barbarous societies, as they are called, every man, it has already been observed, is a warrior. Every man too is in some measure a statesman, and can form a tolerable judgment concerning the interest of the society, and the conduct of those who govern it.... Every man does, or is capable of doing, almost every thing which any other man does, or is capable of doing. Every man has a considerable degree of knowledge, ingenuity, and invention; but scarce any man has a great degree. The degree, however, which is commonly possessed, is generally sufficient for conducting the whole simple business of the society. In a civilized state, on the contrary, though there is little variety in the occupations of the greater part of individuals, there is an almost infinite variety in those of the whole society. These varied occupations present an almost infinite variety of objects to the contemplation of those few, who, being attached to no particular occupation themselves, have leisure and inclination to examine the occupations of other people. The contemplation of so great a variety of objects necessarily exercises their minds in endless comparisons and combinations, and renders their understandings, in an extraordinary degree, both acute and comprehensive. Unless those few, however, happen to be placed in some very particular situations, their great abilities, though honourable to themselves, may contribute very little to the good government or happiness of their society. Notwithstanding the great abilities of those few, all the nobler parts of the human character may be, in a great measure, obliterated and extinguished in the great body of the people....

[The government does not have to provide education for the elite. But it is different for the working class.]

They have little time to spare for education. Their parents can scarce afford to maintain them even in infancy. As soon as they are able to work, they must apply to some trade by which they can earn their subsistence. That trade too is generally so simple and uniform as to give little exercise to the understanding; while, at the same time, their labour is both so constant and so severe, that it leaves them little leisure and less inclination to apply to, or even to think of any thing else.

But though the common people cannot, in any civilized society, be so well instructed as people of some rank and fortune, the most essential parts of education, however, to read, write, and account, can be acquired at so early a period of life, that the greater part even of those who are to be bred to the lowest occupations, have time to acquire them before they can be employed in those occupations. For a very small expense the public can facilitate, can encourage, and can even impose upon almost the whole body of the people, the necessity of acquiring those most essential parts of education.

The public can facilitate this acquisition by establishing in every parish or district a little school, where children may be taught for a reward so moderate, that even a common labourer may afford it; the master being partly, but not wholly paid by the public; because, if he was wholly, or even principally paid by it, he would soon learn to neglect his business. In Scotland the establishment of such parish schools has taught almost the whole common people to read, and a very great proportion of them to write and account. In England the establishment of charity schools has had an effect of the same kind, though not so universally, because the establishment is not so universal. If in those little schools the books, by which the children are taught to read, were a little more instructive than they commonly are; and if, instead of a little smattering of Latin, which the children of the common people are sometimes taught there, and which can scarce ever be of any use to them; they were instructed in the elementary parts of geometry and mechanics, the literary education of this rank of people would perhaps be as complete as it can be. There is scarce a common trade which does not afford some opportunities of applying to it the principles of geometry and mechanics, and which would not therefore gradually exercise and improve the common people in those principles, the necessary introduction to the most sublime as well as to the most useful sciences.

The public can encourage the acquisition of those most essential parts of education by giving small premiums, and little badges of distinction, to the children of the common people who excel in them.

The public can impose upon almost the whole body of the people the necessity of acquiring those most essential parts of education, by obliging every man to undergo an examination or probation in them before he can obtain the freedom in any corporation, or be allowed to set up any trade either in a village or town corporate....

[Workers also should be required to do military and gymnastic exercises in order to develop their "martial spirit."]

Even though the martial spirit of the people were of no use towards the defence of the society, yet to prevent that sort of mental mutilation, deformity, and wretchedness, which cowardice necessarily involves in it, from spreading themselves through the great body of the people, would still deserve the most serious attention of government; in the same manner as it would deserve its most serious attention to prevent a leprosy or any other loathsome and offensive disease, though neither mortal nor dangerous, from spreading itself among them; though, perhaps, no other public good might result from such attention besides the prevention of so great a public evil.

The same thing may be said of the gross ignorance and stupidity which, in a civilized society, seem so frequently to benumb the understandings of all the inferior ranks of people. A man without the proper use of the intellectual faculties of a man, is, if possible, more contemptible than even a coward, and seems to be mutilated and deformed in a still more essential part of the character of human nature. Though the state was to derive no advantage from the instruction of the inferior ranks of people, it would still deserve its attention that they should not be altogether uninstructed. The state, however, derives no inconsiderable advantage from their instruction. The more they are instructed, the less liable they are to the delusions of enthusiasm and superstition, which, among ignorant nations, frequently occasion the most dreadful disorders. An instructed and intelligent people besides, are always more decent and orderly than an ignorant and stupid one. They feel themselves, each individually, more respectable, and more likely to obtain the respect of their lawful superiors and they are therefore more disposed to respect those superiors. They are more disposed to examine, and more capable of seeing through, the interested complaints of faction and sedition, and they are, upon that account, less apt to be misled into any wanton or unnecessary opposition to the measures of government. In free countries, where the safety of government depends very much upon the favourable judgment which the people may form of its conduct, it must surely be of the highest importance that they should not be disposed to judge rashly or capriciously concerning it.

Robert Owen, *A New View of Society*

Robert Owen led a life at the forefront of the social changes inaugurated by the Industrial Revolution. He was both an entrepreneurial manufacturer and a philanthropist. He was convinced that the most profitable business practices were also those that significantly benefited the working class. He was also willing to "put his money where his mouth was," so to speak. Owen assumed management at what he hoped would become a model factory—New Lanark, in Scotland—to put his theories about how to improve the lives of the working class to the test. A New View of Society *draws its recommendations in large part from Owen's experiences in New Lanark.*

Later, Robert Owen emigrated to America, where he continued his experimentation in paternalist and proto-socialist modes of social organization, founding a manufacturing community in New Harmony, Indiana, in 1825, which did not succeed commercially. During his time in England, his energetic impetus was behind many legal reforms, including the 1819 Cotton Mills and Factories Act, which contained legal prohibitions on child labor below the age of nine.

As you read A New View of Society, *your character should consider these questions:*

- *Owen begins with an argument from analogy: just as it is better to spend more to maintain your machines in good working order, the wise merchant should be willing to spend more to maintain his employees in good working order. Would your character agree or disagree? Why?*

- *Robert Owen, in his survey of New Lanark, argues that the ills that beset his workers stem from their bad character. He is confident that providing both moral instruction and superior working conditions can ameliorate these ills and vices among the workers. Would your character share this assessment? How would your experience in Manchester shape your response to this advice?*

Owen spends a great deal of his essay considering the plight of children in the factory system and advocating for his rival method of fostering their wellbeing. Consider his account in light of the debates regarding child welfare in factories. How would his approach address the situation in Manchester? A standard argument against prohibiting children in factories, you will read, is that the family would rather have the income from children working— how would Owen address this argument?

Source: Robert Owen, *A New View of Society, or, Essays on the Principle of the Formation of the Human Character, and the Application of the Principle to Practice* (London: Longman, 1817).

[Address prefixed to the Third Essay.]

To the superintendents of manufactories, and to those individuals generally, who, by giving employment to an aggregated population, may easily adopt the means to form the sentiments and manners of such a population.

Like you, I am a manufacturer for pecuniary profit, but having for many years acted on principles the reverse in many respects of those in which you have been instructed, and having found my procedure beneficial to others and to myself, even in a pecuniary point of view, I am anxious to explain such valuable principles, that you and those under your influence may equally partake of their advantages.

In two Essays, already published, I have developed some of these principles, and in the following pages you will find still more of them explained, with some detail of their application to practice under the peculiar local circumstances in which I took the direction of the New Lanark Mills and Establishment.

By those details you will find that from the commencement of my management I viewed the population, with the mechanism and every other part of the establishment, as a system composed of many parts, and which it was my duty and interest so to combine, as that every hand, as well as every spring, lever, and wheel, should effectually co-operate to produce the greatest pecuniary gain to the proprietors.

Many of you have long experienced in your manufacturing operations the advantages of substantial, well-contrived, and well-executed machinery.

Experience has also shown you the difference of the results between mechanism which is neat, clean, well-arranged, and always in a high state of repair; and that which is allowed to be dirty, in disorder, without the means of preventing unnecessary friction, and which therefore becomes, and works, much out of repair.

In the first case the whole economy and management are good; every operation proceeds with ease, order, and success. In the last, the reverse must follow, and a scene be presented of counteraction, confusion, and dissatisfaction among all the agents and instruments interested or occupied in the general process, which cannot fail to create great loss.

If, then, due care as to the state of your inanimate machines can produce such beneficial results, what may not be expected if you devote equal attention to your vital machines, which are far more wonderfully constructed?

When you shall acquire a right knowledge of these, of their curious mechanism, of their self-adjusting powers; when the proper mainspring shall be applied to their varied movements you will become conscious of their real value, and you will readily be induced to turn your thoughts more frequently from your inanimate to your living machines; you will discover that the latter may be easily trained and directed to procure a large increase of pecuniary gain, while you may also derive from them high and substantial gratification. . . .

I have expended much time and capital upon improvements of the living machinery; and it will soon appear that time and the money so expended in the manufactory at New Lanark, even while such improvements are in progress only, and but half their beneficial effects attained, are now producing a return exceeding fifty per cent, and will shortly create profits equal to cent per cent on the original capital expended in them. . . .

Since the general introduction of inanimate mechanism into British manufactories, man, with few exceptions, has been treated as a secondary and inferior machine; and far more attention has been given to perfect the raw materials of wood and metals than those of body and mind. Give but due reflection to the subject, and you will find that man, even as an instrument for the creation of wealth, may be still greatly improved. . . .

Thus seeing with the clearness of noonday light, thus convinced with the certainty of conviction itself, let us not perpetuate the really unnecessary evils which our present practices inflict on this large proportion of our fellow subjects. Should your pecuniary interests somewhat suffer by adopting the line of conduct now urged, many of you are so wealthy that the expense of founding and continuing at your respective establishments the institutions necessary to improve your animate machines would not be felt, but when you may have ocular demonstration, that, instead of any pecuniary loss, a well-directed attention to form the character and increase the comforts of those who are so entirely at your mercy, will essentially add to your gains, prosperity, and happiness, no reasons, except those founded on ignorance of your self-interest, can in future prevent you from bestowing your chief care on the living machines which you employ. And by so doing you will prevent an accumulation of human misery, of which it is now difficult to form an adequate conception.

That you may be convinced of this most valuable truth, which due reflection will show you is founded on the evidence of unerring facts, is the sincere wish of THE AUTHOR.

First Essay

... According to the last returns under the Population Act, the poor and working classes of Great Britain and Ireland have been found to exceed fifteen millions of persons, or nearly three-fourths of the population of the British Islands.

The characters of these persons are now permitted to be very generally formed without proper guidance or direction, and, in many cases, under circumstances which directly impel them to a course of extreme vice and misery; thus rendering them the worst and most dangerous subjects in the empire; while the far greater part of the remainder of the community are educated upon the most mistaken principles of human nature, such, indeed, as cannot fail to produce a general conduct throughout society, totally unworthy of the character of rational beings. . . .

The chief object of these Essays is to assist and forward investigations of such vital importance to the well-being of this country, and of society in general. . . .

[Owen has discovered principles that he has put into practice for the 2,000–3,000 workers in his factory in New Lanark, Scotland, and which he thinks should be introduced everywhere.] . . . so as to give the greatest sum of happiness to the individual and to mankind. . . . *[These principles will]* . . . ensure, with the fewest possible exceptions, health, strength, and vigour of body . . . *[and]* . . . the happiness of man *[which]* can be erected only on the foundations of health of body and Peace of mind. . . .

Second Essay

[Owen describes the cotton spinning factory in Lanark, Scotland, as it was before he took it over. It was started by a Mr. Dale who was very well intentioned. This was in the late eighteenth century, when cotton mills were first introduced into the area. It was a poor region, and it was only because of the waterfalls necessary to power the factory that it was started there. Finding workers for the factory was difficult.]

. . . It was therefore necessary to collect a new population to supply the infant establishment with labourers. This, however, was no light task; for all the regularly trained Scotch peasantry disdained the idea of working early and late, day after day, within cotton mills. Two modes then only remained of obtaining these labourers; the one, to procure children from the various public charities of the country; and the other, to induce families to settle around the works.

To accommodate the first, a large house was erected, which ultimately contained about 500 children, who were procured chiefly from workhouses and charities in Edinburgh. These children were to be fed, clothed, and educated. . . .

To obtain the second, a village was built; and the houses were let at a low rent to such families as could be induced to accept employment in the mills; but such was the general dislike to that occupation at the time, that, with a few exceptions, only persons destitute of friends, employment, and character, were found willing to try the experiment. . . .

Those who have a practical knowledge of mankind will readily anticipate the character which a population so collected and constituted would acquire. It is therefore scarcely necessary to state, that the community by degrees was formed under these circumstances into a very wretched society. Every man did that which was right in his own eyes, and vice and immorality prevailed to a monstrous extent. The population lived in idleness, in poverty, in almost every kind of crime; consequently, in debt, out of health, and in misery. . . .

The boarding-house containing the children presented a very different scene. The benevolent proprietor spared no expense to give comfort to the poor children. The rooms provided for them were spacious, always clean, and well ventilated; the food was abundant, and of the best quality; the clothes were neat and useful; a surgeon was kept in constant pay, to direct how to prevent or cure disease; and the best instructors which the country afforded were appointed to teach such branches of education as were deemed likely to be useful to children in their situation. Kind and well-disposed persons were appointed to superintend all their proceedings. Nothing, in short, at first sight seemed wanting to render it a most complete charity.

But to defray the expense of these well-devised arrangements, and to support the establishment generally, it was absolutely necessary that the children should be employed within the mills from six o'clock in the morning till seven in the evening, summer and winter; and after these hours their education commenced. The directors of the public charities, from mistaken economy, would not consent to send the children under their care to cotton mills, unless the children were received by the proprietors at the ages of six, seven and eight. And Mr Dale was under the necessity of accepting them at those ages, or of stopping the manufactory which he had commenced.

It is not to be supposed that children so young could remain, with the intervals of meals only, from six in the morning until seven in the evening, in constant employment, on their feet, within cotton mills, and afterwards acquire much proficiency in education. And so it proved; for many of them became dwarfs in body and mind, and some of them were deformed. Their labour through the day and their education at night became so irksome, that numbers of them continually ran away....

Thus Mr Dale's arrangements, and his kind solicitude for the comfort and happiness of these children, were rendered in their ultimate effect almost nugatory.... The error proceeded from the children being sent from the workhouses at an age much too young for employment. They ought to have been detained four years longer, and educated; and then some of the evils which followed would have been prevented.

If such be a true picture, not overcharged, of parish apprentices to our manufacturing system, under the best and most humane regulations, in what colours must it be exhibited under the worst?....

[When Owen took over the factory he applied his new principles to help improve the health and happiness of his workers. The workers resisted the changes at first, but he] ... did not lose his patience, his temper, or his confidence in the certain success of the principles on which he founded his conduct.

These principles ultimately prevailed: the population could not continue to resist a firm well-directed kindness, administering justice to all. They therefore slowly and cautiously began to give him some portion of their confidence; and as this increased, he was enabled more and more to develop his plans for their amelioration. It may with truth be said, that at this period they possessed almost all the vices and very few of the virtues of a social community. Theft and the receipt of stolen goods was their trade, idleness and drunkenness their habit, falsehood and deception their garb, dissensions, civil and religious, their daily practice; they united only in a zealous systematic opposition to their employers.

Here then was a fair field on which to try the efficacy in practice of principles supposed capable of altering any characters. The manager formed his plans accordingly. He spent some time in finding out the full extent of the evil against which he had to contend, and in tracing the true causes which had produced and were continuing those effects. He found that all was distrust, disorder, and disunion; and he wished to introduce confidence, regularity, and harmony. He therefore began to bring forward his various expedients to withdraw the unfavourable circumstances by which they had hitherto been surrounded, and to replace them by others calculated to produce a more happy result. He soon discovered that theft was extended through almost all the ramifications of the community, and the receipt of stolen goods through all the country around. To remedy this evil, not one legal punishment was inflicted, not one individual imprisoned, even for an hour; but checks and other regulations of prevention were introduced; a short plain explanation of the immediate benefits they would derive from a different conduct was inculcated by those instructed for the purpose, who had the best powers of reasoning among themselves. They were at the same time instructed how to direct their industry in legal and useful occupations, by which, without danger or disgrace, they could really earn more than they had previously obtained by dishonest practices. Thus the difficulty of committing the crime was increased, the detection afterwards rendered more easy, the habit of honest industry formed, and the pleasure of good conduct experienced.

Drunkenness was attacked in the same manner; it was discountenanced on every occasion by those who had charge of any department: its destructive and pernicious effects were frequently stated by his own more prudent comrades, at the proper moment when the individual was soberly suffering from the effects of his previous excess; pot- and public-houses were gradually removed from the immediate vicinity of their dwellings; the health and comfort of temperance were made familiar to them; by degrees drunkenness disappeared, and many who were habitual bacchanalians are now conspicuous for undeviating sobriety.

Falsehood and deception met with a similar fate: they were held in disgrace; their practical evils were shortly explained; and every countenance was given to truth and open conduct. The pleasure and substantial advantages derived from the latter soon overcame the impolicy, error, and consequent misery, which the former mode of acting had created.

Dissensions and quarrels were undermined by analogous expedients. When they could not be readily adjusted between the parties themselves, they were stated to the manager; and as in such cases both disputants were usually more or less in the wrong, that wrong was in as few words as possible explained, forgiveness and friendship recommended, and one simple and easily remembered precept inculcated, as the most valuable rule for their whole conduct, and the advantages of which they would experience every moment of their lives; viz.— "That in future they should endeavour to use the same active exertions to make each other happy and comfortable, as they had hitherto done to make each other miserable; and by carrying this short memorandum in their mind, and applying it on all occasions, they would soon render that place a paradise, which, from the most mistaken principle of action, they now made the abode of misery." The experiment was tried: the parties enjoyed the gratification of this new mode of conduct; references rapidly subsided; and now serious differences are scarcely known.

Considerable jealousies also existed on account of one religious sect possessing a decided preference over the others. This was corrected by discontinuing that preference, and by giving a uniform encouragement to those who conducted themselves well among all the various religious persuasions; by recommending the same consideration to be shown to the conscientious opinions of each sect, on the ground that all must believe the particular doctrines which they had been taught, and consequently that all were in that respect upon an equal footing, nor was it possible yet to say which was right or wrong. . . .

The same principles were applied to correct the irregular intercourse of the sexes: such conduct was discountenanced and held in disgrace; fines were levied upon both parties for the use of the support fund of the community. (This fund arose from each individual contributing one sixtieth part of their wages, which, under their management, was applied to support the sick, and injured by accident, and the aged.) But because they had once unfortunately offended against the established laws and customs of society, they were not forced to become vicious, abandoned, and miserable; the door was left open for them to return to the comforts of kind friends and respected acquaintances; and, beyond any previous expectation, the evil became greatly diminished.

The system of receiving apprentices from public charities was abolished; permanent settlers with large families were encouraged, and comfortable houses were built for their accommodation.

The practice of employing children in the mills, of six, seven and eight years of age, was discontinued, and their parents advised to allow them to acquire health and education until they were ten, years old. (It may be remarked, that even this age is too early to keep them at constant employment in manufactories, from six in the morning to seven in the evening. Far better would it be for the children, their parents, and for society, that the first should not commence employment until they attain the age of twelve, when their education might be finished, and their bodies would be more competent to undergo the fatigue and exertions required of them. When parents can be trained to afford this additional time to their children without inconvenience, they will, of course, adopt the practice now recommended.)

The children were taught reading, writing, and arithmetic, during five years, that is, from five to ten, in the village school, without expense to their parents. All the modern improvements in education have been adopted, or are in process of adoption. (To avoid the inconveniences which must ever arise from the introduction of a particular creed into a school, the children are taught to read in such books as inculcate those precepts of the Christian religion, which are common to all denominations.) They may therefore be taught and well-trained before they engage in any regular employment. Another important consideration is, that all their instruction is rendered a pleasure and delight to them; they are much more anxious for the hour of school-time to arrive than to end; they therefore make a rapid progress; and it may be safely asserted, that if they shall not be trained to form such characters as may be most desired, the fault will not proceed from the children; the cause will be in the want of a true knowledge of human nature in those who have the management of them and their parents.

During the period that these changes were going forward, attention was given to the domestic arrangements of the community.

Their houses were rendered more comfortable, their streets were improved, the best provisions were purchased, and sold to them at low rates, yet covering the original expense, and under such regulations as taught them how to proportion their expenditure to their income. Fuel and clothes were obtained for them in the same manner; and no advantage was attempted to be taken of them, or means used to deceive them.

In consequence, their animosity and opposition to the stranger subsided, their full confidence was obtained, and they became satisfied that no evil was intended them; they were convinced that a real desire existed to increase their happiness upon those grounds alone on which it could be permanently increased. All difficulties in the way of future improvement vanished. They were taught to be rational, and they acted rationally. Thus both parties experienced the incalculable advantages of the system which had been adopted. Those employed became industrious, temperate, healthy, faithful to their employers, and kind to each other, while the proprietors were deriving services from their attachment, almost without inspection, far beyond those which could be obtained by any other means than those of mutual confidence and kindness. Such was the effect of these principles on the adults; on those whose previous habits had been as ill-formed as habits could be; and certainly the application of the principles to practice was made under the most unfavourable circumstances....

Let it not, therefore, be longer said that evil or injurious actions cannot be prevented, or that the most rational habits in the rising generation cannot be universally formed. In those characters which now exhibit crime, the fault is obviously not in the individual, but the defects proceed from the system in which the individual was trained. Withdraw those circumstances which tend to create crime in the human character, and crime will not be created. Replace them with such as are calculated to form habits of order, regularity, temperance, industry; and these qualities will be formed. Adopt measures of fair equity and justice, and you will readily acquire the full and complete confidence of the lower orders. Proceed systematically on principles of undeviating persevering kindness, yet retaining and using, with the least possible severity, the means of restraining crime from immediately injuring society; and by degrees even the crimes now existing in the adults will also gradually disappear: for the worst formed disposition, short of incurable insanity, will not long resist a firm, determined, well-directed, persevering kindness. Such a proceeding, whenever practised, will be found the most powerful and effective corrector of crime, and of all injurious and improper habits.

The experiment narrated shows that this is not hypothesis and theory. The principles may be with confidence stated to be universal, and applicable to all times, persons, and circumstances. . . .

These principles, applied to the community at New Lanark, at first under many of the most discouraging circumstances, but persevered in for sixteen years, effected a complete change in the general character of the village, containing upwards of 2,000 inhabitants, and into which, also, there was a constant influx of newcomers. But as the promulgation of new miracles is not for present times, it is not pretended that under such circumstances one and all are become wise and good; or, that they are free from error. But it may be truly stated, that they now constitute a very improved society; that their worst habits are gone, and that their minor ones will soon disappear under a continuance of the application of the same principles; that during the period mentioned, scarcely a legal punishment has been inflicted, or an application been made for parish funds by any individual among them. Drunkenness is not seen in their streets; and the children are taught and trained in the institution for forming their character without any punishment. The community exhibits the general appearance of industry, temperance, comfort, health, and happiness. These are and ever will be the sure and certain effects of the adoption of the principles explained; and these principles, applied with judgement, will effectually reform the most vicious community existing, and train the younger part of it to any character which may be desired; and that, too, much more easily on an extended than on a limited scale. To apply these principles, however, successfully to practice, both a comprehensive and a minute view must be taken of the existing state of the society on which they are intended to operate. The causes of the most prevalent evils must be accurately traced, and those means which appear the most easy and simple should be immediately applied to remove them. . . .

What then remains to prevent such a system from being immediately adopted into national practice? Nothing, surely, but a general destitution of the knowledge of the practice. For with the certain means of preventing crimes, can it be supposed that British legislators, as soon as these means shall be made evident, will longer withhold them from their fellow subjects? No: I am persuaded that neither prince, ministers, parliament, nor any party in church or state, will avow inclination to act on principles of such flagrant injustice. Have they not on many occasions evinced a sincere and ardent desire to ameliorate the condition of the subjects of the empire, when practicable means of amelioration were explained to them, which could be adopted without risking the safety of the state?

[Owen then argues that his system should be extended by the government on a national level.]

. . . *[Owen also argues that the government should create]* "a reserve of employment for the surplus working classes, when the general demand for labour throughout the country is not equal to the full occupation of the whole: that employment to be on useful national objects from which the public may derive advantage equal to the expense which those works may require. . . .

In the next Essay an account will be given of the plans which are in progress at New Lanark for the further comfort and improvement of its inhabitants; and a general practical system be described, by which the same advantages may be gradually introduced among the poor and working classes throughout the United Kingdom.

Third Essay

At the conclusion of the Second Essay, a promise was made that an account should be given of the plans which were in progress at New Lanark for the further improvement of its inhabitants; and that a practical system should be sketched, by which equal advantages might be generally introduced among the poor and working classes throughout the United Kingdom. . . .

[Some improvement had been made at New Lanark, but not enough.]

They had not been taught the most valuable domestic and social habits: such as the most economical method of preparing food; how to arrange their dwellings with neatness, and to keep them always clean and in order; but, what was of infinitely more importance, they had not been instructed how to train their children to form them into valuable members of the community, or to know that principles existed, which, when properly applied to practice from infancy, would ensure from man to man, without chance of failure, a just, open, sincere, and benevolent conduct.

[In order to train the inhabitants in] . . . domestic and social acquirements and habits . . . a building, which may be termed the "new institution," was erected in the centre of the establishment, with an enclosed area before it. The area is intended for a playground for the children of the villagers, from the time they can walk alone until they enter the school. It must be evident to those who have been in the practice of observing children with attention, that much of good or evil is taught to or acquired by a child at a very early period of its life; that much of temper or disposition is correctly or incorrectly formed before he attains his second year; and that many durable impressions are made at the termination of the first twelve or even six months of his existence. The children, therefore, of the uninstructed and ill-instructed, suffer material injury in the formation of their characters during these and the subsequent years of childhood and of youth. It was to prevent, or as much as possible to counteract, these primary evils, to which the poor and working classes are exposed when infants, that the area became part of the New Institution. Into this playground the children are to be received as soon as they can freely walk alone; to be superintended by persons instructed to take charge of them.

Each child . . . on his entrance into the playground, is to be told in language which he can understand, that he is never to injure his playfellows; but that, on the contrary, he is to contribute all in his power to make them happy. This simple precept, when comprehended in all its bearings, and the habits which will arise from its early adoption into practice, if no counteracting principle be forced upon the young mind, will effectually supersede all the errors which have hitherto kept the world in ignorance and misery. So simple a precept, too, will be easily taught, and as easily acquired; for the chief employment of the superintendents will be to prevent any deviation from it in practice. The older children, when they shall have experienced the endless advantages from acting on this principle, will, by their example, soon enforce the practice of it on the young strangers: and the happiness, which the little groups will enjoy from this rational conduct, will ensure its speedy and general and willing adoption. The habit also which they will acquire at this early period of life by continually acting on the principle, will fix it firmly; it will become easy and familiar to them, or, as it is often termed, natural. . . .

[This helps develop the character of children in the right direction.]

The character thus early formed will be as durable as it will be advantageous to the individual and to the community, for by the constitution of our nature, when once the mind fully understands that which is true, the impression of that truth cannot be erased except by mental disease or death; while error must be relinquished at every period of life, whenever it can be made manifest to the mind in which it has been received. This part of the arrangement, therefore, will effect the following purposes: The child will be removed, so far as is at present practicable, from the erroneous treatment of the yet untrained and untaught parents. The parents will be relieved from the loss of time and from the care and anxiety which are now occasioned by attendance on their children from the period when they can go alone to that at which they enter the school. The child will be placed in a situation of safety, where, with its future schoolfellows and companions, it will acquire the best habits and principles, while at mealtimes and at night it will return to the caresses of its parents; and the affections of each are likely to be increased by the separation. The area is also to be a place of meeting for the children from five to ten years of age, previous to and after school-hours, and to serve for a drill-ground, the object of which will be hereafter explained; and a shade will be formed, under which in stormy weather the children may retire for shelter.

These are the important purposes to which a playground attached to a school may be applied. Those who have derived a knowledge of human nature from observation know that man in every situation requires relaxation from his constant and regular occupations, whatever they be: and that if he shall not be provided with or permitted to enjoy innocent and uninjurious amusements, he must and will partake of those which he can obtain, to give him temporary relief from his exertions, although the means of gaining that relief should be most pernicious. For man, irrationally instructed, is ever influenced far more by immediate feelings than by remote considerations.

Those, then, who desire to give mankind the character which it would be for the happiness of all that they should possess, will not fail to make careful provision for their amusement and recreation. The Sabbath was originally so intended. It was instituted to be a day of universal enjoyment and happiness to the human race. It is frequently made, however, from the opposite extremes of error, either a day of superstitious gloom and tyranny over the mind, or of the most destructive intemperance and licentiousness. . . .

[In New Lanark the Sabbath will be] a day of innocent and cheerful recreation to the labouring man . . . *[and a time when workers can]* freely partake, without censure, of the air and exercise to which nature invites them, and which their health demands.

To counteract, in some degree, the inconvenience which arose from the misapplication of the Sabbath, it became necessary to introduce on the other days of the week some innocent amusement and recreation for those whose labours were unceasing, and in winter almost uniform.

In summer, the inhabitants of the village of New Lanark have their gardens and potato grounds to cultivate; they have walks laid out to give them health and the habit of being gratified with the ever-changing scenes of nature—for those scenes afford not only the most economical, but also the most innocent pleasures which man can enjoy; and all men may be easily trained to enjoy them.

In winter the community are deprived of these healthy occupations and amusements; they are employed ten hours and three-quarters every day in the week, except Sunday, and generally every individual continues during that time at the same work: and experience has shown that the average health and spirits of the community are several degrees lower in winter than in summer; and this in part may be fairly attributed to that cause. These considerations suggested the necessity of rooms for innocent amusements and rational recreation.

Many well-intentioned individuals, unaccustomed to witness the conduct of those among the lower orders who have been rationally treated and trained, may fancy such an assemblage will necessarily become a scene of confusion and disorder; instead of which, however, it proceeds with uniform propriety. it is highly favourable to the health, spirits, and dispositions of the individuals so engaged; and if any irregularity should arise, the cause will be solely owing to the parties who attempt to direct the proceedings being deficient in a practical knowledge of human nature. It has been and ever will be found far more easy to lead mankind to virtue, or to rational conduct, by providing them with well-regulated innocent amusements and recreations, than by forcing them to submit to useless restraints, which tend only to create disgust, and often to connect such feelings even with that which is excellent in itself, merely because it has been judiciously associated. . . .

Those who have duly reflected on the nature and extent of the mental movements of the world for the last half-century, must be conscious that great changes are in progress; that man is about to advance another important step towards that degree of intelligence which his natural powers seem capable of attaining. . . .

The last part of the intended arrangement of the New Institution remains yet to be described. This is the Church and its doctrines; and they involve considerations of the highest interest and importance; inasmuch as a knowledge of truth on the subject of religion would permanently establish the happiness of man; for it is the inconsistencies alone, proceeding from the want of this knowledge, which have created, and still create, a great proportion of the miseries which exist in the world. . . .

The principle, then, on which the doctrines taught in the New Institution are proposed to be founded, is, that they shall be in unison with universally revealed facts, which cannot but be true. The following are some of the facts, which, with a view to this part of the undertaking, may be deemed fundamental:

That man is born with a desire to obtain happiness, which desire is the primary cause of all his actions, continues through life, and, in popular language, is called self-interest.

That he is also born with the germs of animal propensities, or the desire to sustain, enjoy, and propagate life; and which desires, as they grow and develop themselves, are termed his natural inclinations.

That he is born likewise with faculties which, in their growth, receive, convey, compare, and become conscious of receiving and comparing ideas.

That the ideas so received, conveyed, compared, and understood, constitute human knowledge, or mind, which acquires strength and maturity with the growth of the individual.

That the desire of happiness in man, the germs of his natural inclinations, and the faculties by which he acquires knowledge, are formed unknown to himself in the womb; and whether perfect or imperfect, they are alone the immediate work of the Creator, and over which the infant and future man have no control.

That these inclinations and faculties are not formed exactly alike in any two individuals; hence the diversity of talents, and the varied impressions called liking and disliking which the same external objects make on different persons, and the lesser varieties which exist among men whose characters have been formed apparently under similar circumstances.

That the knowledge which man receives is derived from the objects around him, and chiefly from the example and instruction of his immediate predecessors.

That this knowledge may be limited or extended, erroneous or true; limited, when the individual receives few, and extended when he receives many ideas; erroneous, when those ideas are inconsistent with the facts which exist around him, and true when they are uniformly consistent with them.

That the misery which he experiences, and the happiness which he enjoys, depend on the kind and degree of knowledge which he receives, and on that which is possessed by those around him. . . .

That when the knowledge which man receives shall be extended to its utmost limit, and true without any mixture of error, then he may and will enjoy all the happiness of which his nature will be capable. . . .

That, in proportion as man's desire of self-happiness, or his self-love, is directed by true knowledge, those actions will abound which are virtuous and beneficial to man. . . .

That when these truths are made evident, every individual will necessarily endeavour to promote the happiness of every other individual within his sphere of action; because he must clearly, and without any doubt, comprehend such conduct to be the essence of self-interest, or the true cause of self-happiness.

Here, then, is a firm foundation on which to erect vital religion, pure and undefiled, and the only one which, without any counteracting evil, can give peace and happiness to man. . . .

There is still another arrangement in contemplation for the community at New Lanark, and without which the establishment will remain incomplete.

It is an expedient to enable the individuals, by their own foresight, prudence, and industry, to secure to themselves in old age a comfortable provision and asylum.

Those now employed at the establishment contribute to a fund which supports them when too ill to work, or superannuated. This fund, however, is not calculated to give them more than a bare existence; and it is surely desirable that, after they have spent nearly half a century in unremitting industry, they should, if possible, enjoy a comfortable independence.

To effect this object, it is intended that in the most pleasant situation near the present village, neat and convenient dwellings should be erected, with gardens attached; that they should be surrounded and sheltered by plantations, through which public walks should be formed; and the whole arranged to give the occupiers the most substantial comforts.

That these dwellings, with the privileges of the public walks, etc., shall become the property of those individuals who, without compulsion, shall subscribe each equitable sums monthly, as, in a given number of years will be equal to the purchase, and to create a fund from which, when these individuals become occupiers of their new residences they may receive weekly, monthly, or quarterly payments, sufficient for their support; the expenses of which may be reduced to a very low rate individually, by arrangements which may be easily formed to supply all their wants with little trouble to themselves; and by their previous instruction they will be enabled to afford the small additional subscription which will be required for these purposes.

This part of the arrangement would always present a prospect of rest, comfort, and happiness to those employed; in consequence, their daily occupations would be performed with more spirit and cheerfulness, and their labour would appear comparatively light and easy. Those still engaged in active operations would, of course, frequently visit their former companions and friends, who, after having spent their years of toil, were in the actual enjoyment of this simple retreat; and from this intercourse each party would naturally derive pleasure. The reflections of each would be most gratifying. The old would rejoice that they had been trained in habits of industry, temperance, and foresight, to enable them to receive and enjoy in their declining years every reasonable comfort which the present state of society will admit; the young and middle-aged, that they were pursuing the same course, and that they had not been trained to waste their money, time, and health, in idleness and intemperance. These and many similar reflections could not fail often to arise in their minds; and those who could look forward with confident hopes to such certain comfort and independence would, in part, enjoy by anticipation these advantages. In short, when this part of the arrangement is well considered, it will be found to be the most important to the community and to the proprietors; indeed, the extensively good effects of it will be experienced in such a variety of ways, that to describe them even below the truth would appear an extravagant exaggeration. They will not, however, prove the less true because mankind are yet ignorant of the practice, and of the principles on which it has been founded.

These, then, are the plans which are in progress or intended for the further improvement of the inhabitants of New Lanark. . . .

In every measure to be introduced at the place in question, for the comfort and happiness of man, the existing errors of the country were always to be considered; and as the establishment belonged to parties whose views were various, it became also necessary to devise means to create pecuniary gains from each improvement, sufficient to satisfy the spirit of commerce.

All, therefore, which has been done for the happiness of this community, which consists of between two and three thousand individuals, is far short of what might have been easily effected in practice had not mankind been previously trained in error. . . .

But to whom can such arrangements be submitted? Not to the mere commercial character, in whose estimation to forsake the path of immediate individual gain would be to show symptoms of a disordered imagination; for the children of commerce have been trained to direct all their faculties to buy cheap and sell dear; and consequently, those who are the most expert and successful in this wise and noble art, are, in the commercial world, deemed to possess foresight and superior acquirements; while such as attempt to improve the moral habits and increase the comforts of those whom they employ, are termed wild enthusiasts.

Nor yet are they to be submitted to the mere men of the law; for these are necessarily trained to endeavour to make wrong appear right, or to involve both in a maze of intricacies, and to legalize injustice.

Nor to mere political leaders or their partisans; for they are embarrassed by the trammels of party, which mislead their judgement, and often constrain them to sacrifice the real well-being of the community and of themselves, to an apparent but most mistaken self-interest. . . .

These principles, therefore, and the practical systems which they recommend, are not to be submitted to the judgement of those who have been trained under, and continue in, any of these unhappy combinations of circumstances. But they are to be submitted to the dispassionate and patient investigation and decision of those individuals of every rank and class and denomination of society, who have become in some degree conscious of the errors in which they exist; who have felt the thick mental darkness by which they are surrounded; who are ardently desirous of discovering and following truth wherever it may lead; and who can perceive the inseparable connection which exists between individual and general, between private and public good!. . . .

Fourth Essay
[Among other things, Owen gives his opinion of the Poor Law in this essay.]

. . . The next measure for the general improvement of the British population should be to revise the laws relative to the poor. For pure and benevolent as, no doubt, were the motives which actuated those with whom the Poor Laws originated, the direct and certain effects of these laws are to injure the poor, and through them, the state, as much almost as they can be injured.

They exhibit the appearance of affording aid to the distressed, while, in reality, they prepare the poor to acquire the worst habits, and to practise every kind of crime. They thus increase the number of the poor and add to their distress. It becomes, therefore, necessary that decisive and effectual measures should be adopted to remove those evils which the existing laws have created.

Benevolence says, that the destitute must not starve; and to this declaration political wisdom readily assents. Yet can that system be right, which compels the industrious, temperate, and comparatively virtuous, to support the ignorant, the idle, and comparatively vicious? Such, however, is the effect of the present British Poor Laws; for they publicly proclaim greater encouragement to idleness, ignorance, extravagance, and intemperance, than to industry and good conduct: and the evils which arise from a system so irrational are hourly experienced, and hourly increasing.

It thus becomes necessary that some counteracting remedy be immediately devised and applied: for, injurious as these laws are, it is obviously impracticable, in the present state of the British population, to annul at once a system to which so large a portion of the people has been taught to look for support.

These laws should be progressively undermined by a system of an opposite nature, and ultimately rendered altogether nugatory.

The proper system to supersede these laws has been in part already explained, but we proceed to unfold it still more. It may be called "A System for the Prevention of Crime, and the Formation of Human Character" and, under an established and well-intentioned government it will be found more efficacious in producing public benefit than any of the laws now in existence.

The fundamental principle on which all these Essays proceed is, that "children collectively may be taught any sentiments and habits" or, in other words, "trained to acquire any character."

It is of importance that this principle should be for ever present in the mind, and that its truth should be established beyond even the shadow of doubt. To the superficial observer it may appear to be an abstract truth of little value; but to the reflecting and accurate reasoner, it will speedily discover itself to be a power which ultimately must destroy the ignorance and consequent prejudices that have accumulated through all preceding ages. . . .

Thomas Malthus, *An Essay on the Principle of Population*

Thomas Malthus was an Anglican vicar and political theorist who flourished around the turn of the eighteenth century. His writings were profoundly influential, and expressed a pessimistic and aristocratic perspective on the problems facing England. In An Essay on the Principle of Population, *Malthus employs a simple demographic model to argue against efforts to ameliorate the condition of the poor, or to achieve a perfect society here on earth.*

Malthus's core argument is that agricultural productivity in England can only grow incrementally, while populations, unless checked in some way, grow at geometric rates. Suppose a population doubles in size over twenty-five years. The second generation, still doubling in size, would start from a base level twice as high as the first. The second generation would create a third generation four times as large as the first, unless something checked that growth. As Malthus sees no decrease in sexual attraction among men and women (there were no effective contraceptives at this time), he reasons that only severe deprivation can check population growth, and therefore that efforts to ameliorate the plight of the poor would only increase the population, bankrupt state coffers, and drive down the market prices of wages, causing mass destitution until population growth recedes.

As you work through his essay, here are some questions for your character to consider:

- *The thrust of the argument in chapter 1 is against the notion that society as a whole can achieve utopian equality—a state where everyone in society has an abundance of time and material resources. Does his account support this conclusion —why or why not?*
- *In chapter 2, Malthus connects his argument with Smith's discussion of the natural and market price of labor. Why does Malthus think that we should treat the labor as a commodity subject to market forces? Would your character agree or disagree? Why?*

Source: Thomas Robert Malthus, *An Essay on the Principle of Population As It Affects the Future Improvement of Society, with Remarks on the Speculations of Mr. Godwin, M. Condorcet, and Other Writers* (London: J. Johnson, 1798).

Preface

. . . It is an obvious truth, which has been taken notice of by many writers, that population must always be kept down to the level of the means of subsistence; but no writer, that the Author recollects, has inquired particularly into the means by which this level is effected: and it is a view of these means, which forms, to his mind, the strongest obstacle in the way to any very great future improvement of society. He hopes it will appear that, in the discussion of this interesting subject, he is actuated solely by a love of truth; and not by any prejudices against any particular set of men, or of opinions. . . .

The view which he has given of human life has a melancholy hue; but he feels conscious, that he has drawn these dark tints, from a conviction that they are really in the picture; and not from a jaundiced eye or an inherent spleen of disposition. . . .

If he should succeed in drawing the attention of more able men, to what he conceives to be the principal difficulty in the way to the improvement of society, and should, in consequence, see this difficulty removed, even in theory, he will gladly retract his present opinions and rejoice in a conviction of his error.

June 7, 1798

Chapter I

. . . The great and unlooked for discoveries that have taken place of late years in natural philosophy; the increasing diffusion of general knowledge from the extension of the art of printing; the ardent and unshackled spirit of inquiry that prevails throughout the lettered and even unlettered world; the new and extraordinary lights that have been thrown on political subjects, which dazzle, and astonish the understanding; and particularly that tremendous phenomenon in the political horizon the French Revolution, which, like a blazing comet, seems destined either to inspire with fresh life and vigour, or to scorch up and destroy the shrinking inhabitants of the earth, have all concurred to lead many able men into the opinion that we were touching on a period big with the most important changes, changes that would in some measure be decisive of the future fate of mankind.

It has been said that the great question is now at issue, whether man shall henceforth start forwards with accelerated velocity towards illimitable, and hitherto unconceived improvement; or be condemned to a perpetual oscillation between happiness and misery, and after every effort remain still at an immeasurable distance from the wished-for goal....

I have read some of the speculations on the perfectibility of man and of society with great pleasure. I have been warmed and delighted with the enchanting picture which they hold forth. I ardently wish for such happy improvements. But I see great, and, to my understanding, unconquerable difficulties in the way to them....

I think I may fairly make two postulata.

First, That food is necessary to the existence of man.

Secondly, That the passion between the sexes is necessary and will remain nearly in its present state.

These two laws, ever since we have had any knowledge of mankind, appear to have been fixed laws of our nature; and, as we have not hitherto seen any alteration in them, we have no right to conclude that they will ever cease to be what they now are, without an immediate act of power in that Being who first arranged the system of the universe; and for the advantage of his creatures, still executes, according to fixed laws, all its various operations.

I do not know that any writer has supposed that on this earth man will ultimately be able to live without food. But Mr. Godwin has conjectured that the passion between the sexes may in time be extinguished. As, however, he calls this part of his work a deviation into the land of conjecture, I will not dwell longer upon it at present than to say that the best arguments for the perfectibility of man, are drawn from a contemplation of the great progress that he has already made from the savage state, and the difficulty of saying where he is to stop. But towards the extinction of the passion between the sexes, no progress whatever has hitherto been made. It appears to exist in as much force at present as it did two thousand or four thousand years ago....

Assuming then, my postulata as granted, I say, that the power of population is indefinitely greater than the power in the earth to produce subsistence for man.

Population, when unchecked, increases in a geometrical ratio. Subsistence increases only in an arithmetical ratio. A slight acquaintance with numbers will shew the immensity of the first power in comparison of the second.

By that law of our nature which makes food necessary to the life of man, the effects of these two unequal powers must be kept equal.

This implies a strong and constantly operating check on population from the difficulty of subsistence. This difficulty must fall somewhere; and must necessarily be severely felt by a large portion of mankind.

Through the animal and vegetable kingdoms, nature has scattered the seeds of life abroad with the most profuse and liberal hand. She has been comparatively sparing in the room and the nourishment necessary to rear them. The germs of existence contained in this spot of earth, with ample food, and ample room to expand in, would fill millions of worlds in the course of a few thousand years. Necessity, that imperious all pervading law of nature, restrains them within the prescribed bounds. The race of plants, and the race of animals shrink under this great restrictive law. And the race of man cannot, by any efforts of reason, escape from it. Among plants and animals its effects are waste of seed, sickness, and premature death. Among mankind, misery and vice. The former, misery, is an absolutely necessary consequence of it. Vice is a highly probable consequence, and we therefore see it abundantly prevail; but it ought not, perhaps, to be called an absolutely necessary consequence. The ordeal of virtue is to resist all temptation to evil.

This natural inequality of the two powers of population, and of production in the earth, and that great law of our nature which must constantly keep their effects equal, form the great difficulty that to me appears insurmountable in the way to the perfectibility of society. All other arguments are of slight and subordinate consideration in comparison of this. I see no way by which man can escape from the weight of this law which pervades all animated nature. No fancied equality, no agrarian regulations in their utmost extent, could remove the pressure of it even for a single century. And it appears, therefore, to be decisive against the possible existence of a society, all the members of which, should live in ease, happiness, and comparative leisure; and feel no anxiety about providing the means of subsistence for themselves and families.

Consequently, if the premises are just, the argument is conclusive against the perfectibility of the mass of mankind. . . .

Chapter II

. . . I said that population, when unchecked, increased in a geometrical ratio; and subsistence for man in an arithmetical ratio.

Let us examine whether this position be just.

I think it will be allowed, that no state has hitherto existed (at least that we have any account of) where the manners were so pure and simple, and the means of subsistence so abundant, that no check whatever has existed to early marriages; among the lower classes, from a fear of not providing well for their families; or among the higher classes, from a fear of lowering their condition in life. Consequently in no state that we have yet known has the power of population been left to exert itself with perfect freedom.

Whether the law of marriage be instituted, or not, the dictate of nature and virtue, seems to be an early attachment to one woman. Supposing a liberty of changing in the case of an unfortunate choice, this liberty would not affect population till it arose to a height greatly vicious; and we are now supposing the existence of a society where vice is scarcely known.

II.5
In a state therefore of great equality and virtue, where pure and simple manners prevailed, and where the means of subsistence were so abundant, that no part of the society could have any fears about providing amply for a family, the power of population being left to exert itself unchecked, the increase of the human species would evidently be much greater than any increase that has been hitherto known.

In the United States of America, where the means of subsistence have been more ample, the manners of the people more pure, and consequently the checks to early marriages fewer, than in any of the modern states of Europe, the population has been found to double itself in twenty-five years.

This ratio of increase, though short of the utmost power of population, yet as the result of actual experience, we will take as our rule; and say,

> That population, when unchecked, goes on doubling itself every twenty-five years or increases in a geometrical ratio.

Let us now take any spot of earth, this Island for instance, and see in what ratio the subsistence it affords can be supposed to increase. We will begin with it under its present state of cultivation.

If I allow that by the best possible policy, by breaking up more land and by great encouragements to agriculture, the produce of this Island may be doubled in the first twenty-five years, I think it will be allowing as much as any person can well demand.

II.10
In the next twenty-five years, it is impossible to suppose that the produce could be quadrupled. It would be contrary to all our knowledge of the qualities of land. The very utmost that we can conceive, is, that the increase in the second twenty-five years might equal the present produce. Let us then take this for our rule, though certainly far beyond the truth; and allow that, by great exertion, the whole produce of the Island might be increased every twenty-five years, by a quantity of subsistence equal to what it

at present produces. The most enthusiastic speculator cannot suppose a greater increase than this. In a few centuries it would make every acre of land in the Island like a garden.

Yet this ratio of increase is evidently arithmetical.

It may be fairly said, therefore, that the means of subsistence increase in an arithmetical ratio.

Let us now bring the effects of these two ratios together.

The population of the Island is computed to be about seven Millions; and we will suppose the present produce equal to the support of such a number. In the first twenty-five years the population would be fourteen millions; and the food being also doubled, the means of subsistence would be equal to this increase. In the next twenty-five years the population would be twenty-eight millions; and the means of subsistence only equal to the support of twenty-one millions. In the next period, the population would be fifty-six millions, and the means of subsistence just sufficient for half that number. And at the conclusion of the first century the population would be one hundred and twelve millions and the means of subsistence only equal to the support of thirty-five millions; which would leave a population of seventy-seven millions totally unprovided for.

A great emigration necessarily implies unhappiness of some kind or other in the country that is deserted. For few persons will leave their families, connections, friends, and native land, to seek a settlement in untried foreign climes, without some strong subsisting causes of uneasiness where they are, or the hope of some great advantages in the place to which they are going.

But to make the argument more general and less interrupted by the partial views of emigration, let us take the whole earth, instead of one spot, and suppose that the restraints to population were universally removed. If the subsistence for man that the earth affords was to be increased every twenty-five years by a quantity equal to what the whole world at present produces; this would allow the power of production in the earth to be absolutely unlimited, and its ratio of increase much greater than we can conceive that any possible exertions of mankind could make it.

Taking the population of the world at any number, a thousand millions, for instance, the human species would increase in the ratio of—1, 2, 4, 8, 16, 32, 64, 128, 256, 512, &c. and subsistence as—1, 2, 3, 4, 5, 6, 7, 8, 9, 10, &c. In two centuries and a quarter, the population would be to the means of subsistence as 512 to 10: in three centuries as 4096 to 13; and in two thousand years the difference would be almost incalculable, though the produce in that time would have increased to an immense extent.

No limits whatever are placed to the productions of the earth; they may increase for ever and be greater than any assignable quantity; yet still the power of population being a power of a superior order, the increase of the human species can only be kept commensurate to the increase of the means of subsistence, by the constant operation of the strong law of necessity acting as a check upon the greater power.

The effects of this check remain now to be considered.

Among plants and animals the view of the subject is simple. They are all impelled by a powerful instinct to the increase of their species; and this instinct is interrupted by no reasoning, or doubts about providing for their offspring. Wherever therefore there is liberty, the power of increase is exerted; and the superabundant effects are repressed afterwards by want of room and nourishment, which is common to animals and plants; and among animals by becoming the prey of others.

The effects of this check on man are more complicated.

Impelled to the increase of his species by an equally powerful instinct, reason interrupts his career, and asks him whether he may not bring beings into the world, for whom he cannot provide the means of subsistence. In a state of equality, this would be the simple question. In the present state of society, other considerations occur. Will he not lower his rank in life? Will he not subject himself to greater difficulties than he at present feels? Will he not be obliged to labour harder? and if he has a large family, will his utmost exertions enable him to support them? May he not see his offspring in rags and misery, and clamouring for bread that he cannot give them? And may he not be reduced to the grating necessity of forfeiting his independence, and of being obliged to the sparing hand of charity for support?

These considerations are calculated to prevent, and certainly do prevent, a very great number in all civilized nations from pursuing the dictate of nature in an early attachment to one woman. And this restraint almost necessarily, though not absolutely so, produces vice. Yet in all societies, even those that are most vicious, the tendency to a virtuous attachment is so strong that there is a constant effort towards an increase of population. This constant effort as constantly tends to subject the lower classes of the society to distress and to prevent any great permanent amelioration of their condition.

The way in which these effects are produced seems to be this.

We will suppose the means of subsistence in any country just equal to the easy support of its inhabitants. The constant effort towards population, which is found to act even in the most vicious societies, increases the number of people before the means of subsistence are increased. The food therefore which before supported seven millions, must now be divided among seven millions and a half or eight millions. The poor consequently must live much worse, and many of them be reduced to severe distress. The number of labourers also being above the proportion of the work in the market, the price of labour must tend toward a decrease; while the price of provisions would at the same time tend to rise. The labourer therefore must work harder to earn the same as he did before. During this season of distress, the discouragements to marriage, and the difficulty of rearing a family are so great, that population is at a stand. In the mean time the cheapness of labour, the plenty of labourers, and the necessity of an increased industry amongst them, encourage cultivators to employ more labour upon their land; to turn up fresh soil, and to manure and improve more completely what is already in tillage; till ultimately the means of subsistence become in the same proportion to the population as at the period from which we set out. The situation of the labourer being then again tolerably comfortable, the restraints to population are in some degree loosened; and the same retrograde and progressive movements with respect to happiness are repeated.

This sort of oscillation will not be remarked by superficial Observers; and it may be difficult even for the most penetrating mind to calculate its periods. Yet that in all old states some such vibration does exist; though from various transverse causes, in a much less marked, and in a much more irregular manner than I have described it, no reflecting man who considers the subject deeply can well doubt. . . .

It very rarely happens that the nominal price of labour universally falls; but we well know that it frequently remains the same, while the nominal price of provisions has been gradually increasing. This is, in effect, a real fall in the price of labour; and during this period, the condition of the lower orders of the community must gradually grow worse and worse. But the farmers and capitalists are growing rich from the real cheapness of labour. Their increased capitals enable them to employ a greater number of men. Work therefore may be plentiful; and the price of labour would consequently rise. But the want of freedom in the market of labour, which occurs more or less in all communities, either from parish laws, or the more general cause of the facility of combination among the rich, and its difficulty among the poor, operates to prevent the price of labour from rising at the natural period, and keeps it down some time longer; perhaps till a year of scarcity, when the clamour is too loud, and the necessity too apparent to be resisted.

The true cause of the advance in the price of labour is thus Concealed; and the rich affect to grant it as an act of compassion and favour to the poor, in consideration of a year of scarcity; and when plenty returns, indulge themselves in the most unreasonable of all complaints, that the price does not again fall; when a little reflection would shew them that it must have risen long before but from an unjust conspiracy of their own.

But though the rich by unfair combinations, contribute frequently to prolong a season of distress among the poor; yet no possible form of society could prevent the almost constant action of misery upon a great part of mankind, if in a state of inequality, and upon all, if all were equal.

The theory on which the truth of this position depends appears to me so extremely clear; that I feel at a loss to conjecture what part of it can be denied.

That population cannot increase without the means of Subsistence, is a proposition so evident, that it needs no illustration.

That population does invariably increase, where there are the means of subsistence, the history of every people that have ever existed will abundantly prove.

And, that the superior power of population cannot be checked, without producing misery or vice, the ample portion of these too bitter ingredients in the cup of human life, and the continuance of the physical causes that seem to have produced them, bear too convincing a testimony.

But in order more fully to ascertain the validity of these three propositions, let us examine the different states in which mankind have been known to exist. Even a cursory review will, I think, be sufficient to convince us, that these propositions are incontrovertible truths. . . .

Proposals to Help Handloom Weavers

These proposals come from a later period than when this game is set, but the issues were debated as early as 1808, when Parliament considered petitions of cotton weavers. See Great Britain, Parliament, House of Commons. Committee on Petitions of Several Cotton Manufacturers and Journeymen Cotton Weavers, Report *(London: 1808).*

Here are some questions to consider regarding arguments for and against establishing a legal minimum wage, as well as arguments for and against raising taxes on machines. The discussion of taxation here will be helpful when read in conjunction with later texts on the question of taxes in early nineteenth-century England.

- *According to the first author in favor of the minimum wage, how is labor different from other commodities? Would your character agree or disagree with the idea that these differences establish the need for a minimum wage? Why?*
- *What benefits does the author foresee for England as a whole, and not merely the working classes, if a minimum wage is adopted? How would your character evaluate this argument?*
- *What difficulties do manufacturers foresee if they are forced to pay a minimum wage? What alternatives does the opponent of minimum wages think that weavers should pursue?*

Another proposal would impose a tax on the use of machinery in factories. (Here's a question to test your understanding of Smith's theory of supply and demand: How would a tax on machines lower the effective demand for machines?)

What are the arguments advanced in favor of a tax on machines and against this tax? How do you think Smith would reply (recall his insights regarding how inventions and the division of labor increase productivity) or Owen (who wanted to use the productive power of factories to fund his ameliorative programs)? How would a Luddite respond?

Source: Great Britain, Select Committees on Hand-Loom Weavers' Petitions, 1834–1835, *Analysis of the Evidence Taken Before the Select Committees on Hand-Loom Weavers' Petitions (1834–1835): Ordered, by the House of Commons, to Be Printed, 10 August 1835* (London, 1835).

Part I. Remedies for the Distress.
[1. Minimum wage law]

. . . THE leading remedy suggested for the relief of the distressed condition of the hand-loom weavers, is the establishment, by legislative enactment, of a minimum rate of wages throughout the United Kingdom, below which no manufacturer shall be permitted to descend. . . . I commence laying before the Committee a faithful analyzation of the whole of the evidence pro and con on the subject. . . .

[Arguments in favor of a minimum wage law:]
Necessity for protecting labour.—It is asserted that labour, like every other marketable commodity, should find its own level. . . . *[But]* Labour is always carried to market by those who have nothing else to keep or to sell, and who, therefore, must part with it immediately, whether the price pleases them or not, or suffer privations, most likely severe want: labour is always purchased by capitalists, who can abstain from purchasing until they can have it on their own terms, without suffering either privation or want. That, Sir, I presume to call one great distinction essential to the very existence of labour and capital. Another is, that all kinds of commodities . . . having assumed a tangible, visible, substantial form, can be retained if an inadequate price be offered for them, and the fortune of another day, week, month, year, be tried for them, as they will generally keep for some time without suffering much damage; many of them will keep any length of time—without any damage at all. But how fares labour in this respect? the labour which I ought to perform or might perform this week, if I, in imitation of the capitalist, refuse to part with it, that is, refuse to perform it, because an inadequate price is offered me for it, can I bottle it? can I lay it up in salt? In what way can I store it, that, in imitation of the capitalist, I may either get the same price for it, or a higher or lower price, as the case may be, at some future period? That I presume to call another essential difference inseparable from the very nature of labour and capital These two distinctions between the nature of labour and capital (viz. that labour is always sold by the poor, and always bought by the rich, and that labour cannot by any possibility be stored, but must be every instant sold or every instant lost), are sufficient to convince me that labour and capital can never with justice be subjected to the same laws: and I must continue to be of that opinion, until somebody convinces me that these distinctions do not exist, or that their existence forms no real difference.

I do not wish to withdraw what protection capital has, but I do say that labour ought not to be left unprotected. We see the shopkeeper protected by the licence of the hawker, and we see the land protected by the corn law, that both the land and the shopkeeper may pay their taxes; but the labourer, while he pays his taxes, is altogether unprotected. What the weavers seek is, that they may have that protection extended to them which is extended to other parties in the State.

The remedy required appears to be, to create a competition amongst manufacturers to give good wages, by protecting those who are disposed to do so, placing the controlling power of regulating wages in their hands, instead of allowing that power to be, as it now is, in the hands of those who generally take the lead in reducing wages. A plan of taking periodical averages of wages paid for all sorts of hand-loom weaving, on which to found a minimum, has been suggested to effect this object, as preferable to boards of trade composed of masters and men. The former plan has been discussed amongst the weavers since the Report of the Committee was framed, and has met pretty generally with their approval. . . .

Mr. John Makin, an intelligent manufacturer, says, wages form so small a portion of all the branches of manufacture, that they might be raised, so as to place the weaver in far better circumstances, without in any way injuring the trade; the wages of the branch in which I am engaged form only about three eighths of the cost of the goods. Looking at what has fallen under my own observation, I have seen some particular kinds of cloth, owing to the rise of cloth, advancing from 4 s. 6 d. to 5s. 10d. within the last 20 months; also, considering the small portion of the cost of the article which the wages now form, and considering that the purchaser would have confidence in the price of goods remaining where they are, or ascending, I consider that they might be raised equivalent to putting it in the power of the manufacturer to pay 25 percent, more wages than he does now. In the last December but one there was a certain species of cloth, which was sold at 4 s. 6 d. the piece; and in a very short time they have risen to 5 s. 10 d., all owing to the rise of cotton; and the price is obtained as readily now as when it was 4 s. 6 d.; there is sufficient wealth in this country to give employment, at adequate wages, to all branches of industry. The consumers in the country could well afford to pay an additional increase in the price, of a gown; if the wages were raised 25 per cent, and a lady had six yards of muslin in a gown at 1 s. a yard, it would not advance the price of that gown 1 s. I do not believe that any lady would object to give 1 s. more if she thought it would contribute to give comfort and happiness to thousands; I am sure she would be very happy to do it.

This increase of wages, by raising, in the same degree, all those articles which are in use among the weaving population, would tend to increase that portion of the revenue which is collected upon the necessaries of life; they would use more tea, more sugar, more tobacco, more malt, more of every-thing that is taxable; therefore this 25 per cent would render the people not only more willing, but more able, to bear that large and oppressive taxation which the former witness stated to be the great grievance pressing upon the industrious classes; I know that whilst the wages were good, the people were excessively loyal; and they would laud that government to the skies which would improve their condition, be it ever so little. I do not think they would ever think about the fundholder if they got tea and sugar, and coffee and tobacco, and the other comforts of life; they would not even grudge the Civil List and the Pension List. . . . The wife or the daughter of a fundholder, or of a judge, would do well to consent to pay this additional shilling upon the gown with which they are clothed; I believe they would not only consent to it, but if the price of the article were raised it would be more esteemed. It has now become low to a proverb, and despicable on that account.

The landowner, who has got a contingent prohibition to the introduction of corn in his favour, ought to be willing that his wife and daughter should pay 1 s. more for their gowns than they do at present. A minimum rate for the price of weaving would, I am sure, have a tendency to bring about these desirable changes. . . .

Mr. Oastler . . . I would limit the production, by limiting the hours of labour. I would shorten the hours of labour. I would give to the Boards of Trade the power of regulating, that they should work so many hours a day. You would not find the men working so many hours unless they were obliged to do it; it is absolute starvation that compels them to work so long hours; they want to have time with their wives and children; they want to have a little employment for their minds, and so on; some of them are extremely intelligent men, and they do not want to be engaged from morning till night. If there was a law to enable them to get a decent living by eight or ten hours' labour, you would not find them thronging the factories and the looms so long as they do. I dare say, however, that if the proper minimum price of labour to the weaver was fixed, that would secure the proper regulation of the time of labour, without having recourse to fixing the number of hours the weavers shall work, because they do not work for the love of work; they want to eat. If the wages were calculated, so that a man could earn a good living in eight or ten hours, I have no doubt that there would be work for the best weavers and for the others also. I say that these men do not like to sit in the looms all the day long; they would like to enjoy themselves as reasonable beings; they feel that it is degrading to them to be treated as they have been, like animals. I assure you they feel no pleasure in eternally pushing the shuttle. They work long hours because they are forced, and thus force others to do the same, viz. more work for less wages. I think in regard to those places which we call factories, where there are large engines, a law might very easily be made, and could not be at all evaded, that those engines should run so many hours in the day and no longer. But I was asked respecting the weavers; my opinion is, that if a minimum of wages were established for those weavers, their natural wish to get out of the looms would pretty nearly regulate the time.

It is the business and duty of Government to establish some sort of system by which every honest, industrious labourer can live in comfort, and this mode of doing it which is now spoken of does appear to me to be the best mode which Government can possibly propose; they ought to give protection to labour; I consider it the first essential duty of Government to do it, and I do not think that the Government can claim on any ground the allegiance of the operatives, when they see that capital and property are protected, and their labour is left to chance....

John Lennon, a weaver at Preston, declares, that a minimum rate of wages, by bringing up those bad-paying masters to the price given by the good ones, would immediately benefit the unfortunate weavers who are in debt, and enable them to get such reward for their labour as would materially tend to enable them to free themselves from the state in which they are, and to return to their former independent condition. If the weavers had an advance of a penny a yard, it would gradually grow to the benefit of the masters and the men, for the consumers could pay the difference, provided it did not injure our foreign trade. And if a farmer's wife or a lady goes into a shop to purchase an article, she will not leave behind her an article she fancies for the sake of a penny a yard; and if that was added to the weaver's earnings, we should scarcely look for more, for it would increase itself. I think less than a penny a yard would be of little service, or about 2 s. a week. It is not the consumers that wish to keep the weavers in that state of distress, if they could bring them out by paying a small additional price for goods; no, they would be delighted at it....

[Arguments against a minimum wage law:]

The opposition to the measure, in the evidence adduced, is principally developed in leading questions to the witnesses, and answered negatively or affirmatively; but one gentleman, Robert Gardner, Esq.,... says, Yet, speaking for myself, I am very much averse to fettering trade at all, or interfering; because the supply and demand will always regulate wages. Some time ago, perhaps 10 or 15 years ago, we had a great deal of disturbance in Lancashire, principally among the weavers, and there were several meetings; and at that time I was very anxious we should have had a committee of masters to regulate wages; and my only reason (and I stated it to several) for not pursuing it was, I thought it would do mischief, because they would fix them so high as to injure the demand; because no manufacturer of any respectability, if the power lay in his hands, would ever think of fixing the wages at such a price that the weaver could not live well. And I have thought myself that it is necessary for the existence of our trade that wages should not be high. I find that in most parts of Germany they can lay down yarn at an expense of 2d. per pound; and in any part of Switzerland, and the more distant parts of Germany and Saxony, it can be laid down at 3d. per pound on the Manchester price. ... Mr. Gardner continues: Under no circumstances should I be friendly to the regulating of wages: and under the present circumstances, when we are pressed so much with foreign competition, for we ourselves, as an individual house, are supplanted in the Dutch market, in the Belgium market, in the German market for several articles of our manufactory, and also in the American market, with coloured goods, to a considerable extent; and when, as I stated before, yarn can be laid down in the foreign market at from 2d. to 3d. on the Manchester price, and a labourer can support his family for one-third or one quarter of what he can do in this country, I think that is sufficient to show us the great competition; and then power-looms, you are well aware, are encroaching very much indeed, except they were to lay a tax on power-looms. With respect to the comforts of the weavers, my decided opinion is, that wages cannot advance, and therefore I think he is the weaver's friend who would lead the weaver to get any other employment. I asked the question, why is there at least 25 per cent of looms standing in Bolton, and only 10 per cent in different parts? my manager replies, because there is a greater variety of other work at Bolton; for instance, he named two or three of our own weavers, one had got employment at Mr. Hicks's new foundry; and that two or three children, that were weaving for us, were also going to other employments, perhaps in a machine establishment; and that is the way he accounted for a greater proportion of looms standing in Bolton than in country districts, where there is a less variety of other employments. Those who worked these looms have got other employment; formerly a great many of the looms were worked by journeymen, and there is hardly such a thing as a journeyman to be found anywhere now because those men, instead of going into a poor man's house and working on his looms, go into a power-loom mill, where they have more comfort, and perhaps earn 3 s. or 4 s. a week more. . . .

A rise . . . [in wages] would, in my decided opinion, act very injuriously both to the manufacturer and to the country at large. I object to the principle on account of the foreign competition; not less than a dozen individuals have come from Holland, from Belgium and from Germany, this year, that have been in the habit of buying of us regularly certain descriptions of goods; this spring, they said, "We cannot now buy so and so, the Swiss people cut you out; we cannot buy so and so, the Saxon cuts you out"; and therefore I adduce these reasons to show that an advance of wages would drive away that portion of manufacture that is still left. Even supposing the higher paying masters had the power to alter that minimum, I would have the same objection to the Act; because respectable masters would always be governed and influenced by what weavers could live upon, and they must, of course, as men of common feeling, raise the wages considerably, and would, I have no doubt. If this thing had been in operation, perhaps 20 years ago, I think it might have been beneficial to the country, because it would have prevented wages going down so low as what they have done; and I am of opinion, that if wages had never gone so low as they have been, the protecting duties of foreign countries would never have been so high; that is my decided opinion. The countries to which I refer, when speaking of the protecting duties, are the Americans, Germans and different foreign countries. I am alluding now to 20 or 25 years ago. My objection to this system is because of the foreign competition on the one hand, and the great increase of power-looms on the other. . . . Men would be influenced, as I said before, in fixing the wages, considering what a man could support his family for; and it would raise it so high, that it would at once cut off that branch of our foreign business with Europe. At this time, I beg to state distinctly, the hand-loom manufacturers are in a very bad situation; most of them have very heavy stocks, and most of them at this time are curtailing manufacture, in consequence of not being enabled to make sales; and I have no doubt, in the course of a very short time, there will be a general reduction of wages, in consequence, of the hand-loom weavers, partly from foreign competition and partly from the power-loom. . . . I do, therefore, decidedly entertain this opinion, that if the controlling power of regulating wages was placed with the high paying masters, . . . that such a regulation and the placing of power in such hands would be injurious to the country. It would work injuriously, as it would tend to the great and more rapid increase of power-looms. There is such a spirit of enterprize, and such great improvements taking place in power-looms, that I have no doubt of their capabilities, to say the least of it, to weave one-half of the different fabrics that are now woven by hand; and they not only make them cheaper, but better than what they can be made at the present wages. . . . I wish clearly to be understood; I am not opposing an advance of wages on the principle of opposing it; but I say we are so much pressed by power-loom and by foreign competition, that one-half of those weavers that are now employed must be dispersed, and find other employment within seven years from this time. An advance, we will say, for instance, from 6 s. 6 d. to 7 s. by the high paying master, would be injurious; because every advance would tend to give our competitors power over us, and every advance of wages would tend to the great increase of power-looms. . . .

[Arguments for taxation on machinery:]

It has been shown . . . that machinery is one of alleged causes of the distress existing, and it is therefore proposed to regulate its application by taxing it, and thus make it contribute to the revenue as well as the mere animal machine.

The natural effects of the improvements in machinery should be to reduce the time of labour, and place the work-people in better circumstances, if it was rightly directed; a direction might be given to those improvements that would produce that effect; and that improvement, instead of producing many evils, might be converted into a blessing. Machinery properly used, instead of making men into slaves, would make them comfortable; but every man who works on a machine he is worse than the West Indian negro that the Government have given so humanely his liberty.

The hand-loom weavers of Leeds remark, Surely relief can be afforded to 800,000 hand-loom weavers. Five years ago we petitioned the House of Commons for a tax on power-looms; our opinion at that time was, that if the duty then paid on printed goods was taken off, which amounted at that time to 500,000 *l.* *[pounds]*, sterling, and a duty of as much value laid on power-looms, it would be the means of throwing more work into the possession of the hand-loom weavers, and by so doing would give them the means of maintaining themselves and families without encumbering the parish, and the goods would be no dearer to the consumer by so doing. And if the taxes were now taken off soap, oil, wool, cotton, dye-wares, and all other raw materials used in the manufactures, and laid on power-looms, it would have the same effect. It is the duty of the Legislature to protect the weak from being oppressed by the strong; we are the weak, but the power-loom is strong; it takes from us our labour, but it does not contribute its share towards relieving the poverty it creates. It ought to be made more available to local rates, but it unfortunately happens that the persons who have the power to lay the rates are mostly interested in machinery, and, either through custom or interest, consider nothing but the value of the buildings.

Those opposed to taxing machinery, say, if you tax it, it will stop all further improvement, and it will stop machinery; there would be equally as much reason in saying, if bricks be taxed there would be no more brick houses built; or if candles be taxed, that there would be none used, but we should sit in the dark. Again, it is argued, if you tax machinery it will destroy our foreign trade. That was done by the corn laws; in the year 1816, before power-looms became so numerous, there was exported from this country to foreign parts, woollen goods to the value of 5,586,364 *l.*, 9s.9d., and in six years after the passing of the corn laws, which was in the year 1841, the value of British woollens exported to foreign parts was reduced to 4,363,980 *l.* 15s. 3d., which is a decrease of 1,222,383 *l.* 14s. 6 d. in the export of British woollens alone. In 1816 there was exported to the United States of America British woollens to the value of 2,241,510 *l.* 13s. 11 d., and in 10 years after, our exports to the same place were only 1,220,834 *l.* 19 s. 5 d. If we had taken their corn in exchange for our goods, the reverse would have been the consequence; and notwithstanding 20 years of profound peace, and all the machinery brought into play, our foreign trade has never regained in value what it was in the years 1816, 1817 and 1818.

We are of opinion that unless a tax be laid on power-looms of 25 per cent, of the wages we receive at present for the same description of work, we shall in a little time be compelled to work for the same wages they do. With our present wages, when in full employment, we can scarcely get the necessaries let alone the comforts of domestic life. What will be our situation then? It cannot be contemplated without horror. Mr. James Green says, in his report to us, that the hand-loom weavers think that their labour ought to be protected as well as the farmer's corn, by a tax on power-looms; we are of the same opinion. . . .

[Arguments against taxation on machinery:]

William Longson says. . . . With reference to the propriety of taxing the power-loom part of machinery . . . the tax would fall upon the operatives engaged upon the machines in the first place; I see no means of evading it. My reasons for entertaining that opinion are, because the employers fix the rate of wages. . . . I should think that eventually it would be declared . . . that in consequence of such a tax they could only give a certain rate, and would diminish in proportion to that. The operation of a tax upon machinery, whereby the cost of production would be enhanced, would either raise the price of the article manufactured in that way, or, in that case, fall upon the operator working at the machine; in my opinion it would fall upon the operative at the machine. . . . because it being a disagreeable trade, the weavers, if their circumstances were raised and improved, would go back to the hand-loom if he tried to reduce their wages?. . . . In the power-loom weaving department there are many who are hand-loom weavers, and of course various inducements to reduce wages exist, and the necessity of standing during the whole employment would be one very great inducement that might cause them to return to manual labour. . . . However, morally speaking, any two persons who are in competition with each other as labourers ought to pay exactly the same taxation. . . . In my opinion, the laying a tax upon the power-loom would have the effect of reducing the wages of the power-loom weavers; it would immediately fall upon the power-loom weavers. . . .

Three or 400 years ago there were no printed books; it would not have been for the benefit of society to have laid a tax upon printing presses to such an extent that all books should have continued to have been written instead of printed. I say decidedly not; but the persons who were scribes at that time had been taken by society, and were convenient to society; they had learned that occupation; and would it not have been fair and reasonable that they should have been remunerated? might they not have given them pensions because they had learnt a trade to serve you? and what right have you to tell them to go into the grave, or into oblivion, or whatever you call it? If you do not think proper to tax the printing press, the society who are benefited by it ought to compensate the scribes. If you lay hold of a boy and make him a blacksmith to shoe horses; if you get a machine to do it, you have deprived mankind of the benefit of that man's labour; you have confined him to that kind of employment, and why not remunerate him? you may call it a tax or a pension. I would only compensate him till he was able to find some other mode of employment. I would stop no improvement. If I could produce all the objects of human desire by a machine, I should think it my duty to God and to man, to appeal to the State, and say, how are these men to be lodged and clothed, whose labour must be superseded by the introduction of my machine. Whatever may have been the influence and effect of machinery acting in competition with the labourers, who pay heavy taxes to the State, the best mode by which relief can be given to the hand-loom weavers, is some regulation which will prevent the avaricious or improvident manufacturer from reducing the price of weaving, not with reference to the demand of the consumers, not with reference to the price that his commodity could be sold at, but with reference alone to the misery and wretchedness of the weavers, which compel them to take any wages that any person may choose to give them. . . .

Mr. Myerscough observes, With reference to the cloth made by the power-loom contributing in the same degree to the revenue as the cloth made by hand-loom, I should be very sorry for myself to see any tax put on power, believing as I do, that it will not have the slightest effect to improve the condition of the hand-loom weaver. I should like to see the hand-loom weavers relieved from many of the taxes now collected. I should doubt that taxing the loom would raise the price of power-loom cloth, it might certainly interfere very materially with the manufacturer of power-loom cloth, but I am not aware that it would very materially raise the price, so as to afford the hand-loom weavers any benefit.

Mr. Oastler is of opinion that it is very injurious to tax machinery, but I think it is still more injurious to allow machinery to pine the operatives to death. I think it would be a dangerous experiment to tax machinery: but still if nothing else can be done so as to increase the price of labour, that must be done, because the labourer will not submit to be starved much longer; but I do not think that taxing machinery would be a wise legislative measure; I think it would be taxed sufficiently if the time of its work were shortened; I think that would be the best way of taxing them; I think that you ought to have only one tax, but I do not want to increase taxes; I do not think any good will be done by increasing taxes, but if nothing else can be done to enable the labourer to get a living, I see no other way but to resort to taxing machinery. I certainly should say, that if Parliament refuse to shorten the hours of labour, there is no other way of enabling the hand-loom weavers to live but by fixing a tax upon the produce of the power-loom, so as to enable them to compete with the power-looms; but I prefer the shortening of the hours of labour to those new taxations. . . .

[*Arguments for the reduction in taxation on the necessities of life:*]

In opposition to the project for taxing machinery, it is proposed by several witnesses to remove as much as possible the taxes which press on the productive industry of the country:—

Mr. Myerscough says, I think no board of trade can relieve the weavers as long as the pressure of taxation exists upon them. I know, before the duty on salt was taken off, it cost me, when I got a pig salted at my relations, 5s.; now if any tax-gatherer should come and cut 10 lbs. of bacon off the side a man had in his house, he would look very queer at him. The malt tax being taken off would relieve them. The mode of relief I would suggest would be to reduce the taxes from 50,000,000 *l.* to 10,000,000 *l.* or 12,000,000 *l.* In case that cannot be effected, I know of no other mode of relief which presents itself to my mind that is practicable.

Mr. Oastler remarks, I think that if you take away all indirect taxes, and lay a tax upon property and capital, making capital first pay up its arrears of taxation, you would set matters right. An enormous mass of capital has not a single sixpence of taxes to pay anywhere.

Mr. Montgomery Martin says, I would consider it a much more healthy state of society if some means could be adopted to raise the value of labour in this country, and thereby increase the power of consumption among the labouring people to a degree which would render us much less dependent on foreign trade than we now are. I do not, however, think that the wages of labour can be raised by legislative enactment; they must regulate themselves, not only in England, but in contiguous countries, as long as we are at peace, and instead of endeavouring to raise the wages of labour, I would endeavour to take off those burthens which press upon the British labourer, and to enable him to compete fairly with the foreigner.

Capital, like manure, ought to be equally distributed, and not collected in great masses, if the greatest possible public good is contemplated in the financial system of the country. I recommend property tax, as having a tendency to correct that mal-administration in the affairs of this country, which appears by the increase of probate duties to have led to masses of great wealth, while the great mass of the population have been reduced to great misery and distress. I think the removal of taxes which affect the industry of the country, and the levying of a tax on the wealth of the country, levied according to the means of each portion of the community, would have saved the necessity of this Committee, and those further labours to which this will lead with respect to the productive classes. . . .

I have no doubt from the consideration given to this subject, that if the taxes on the articles of consumption named were very materially reduced it would not only give relief to the labouring classes, but it would very materially tend to improve and extend the trade of the country. . . . I think the relief of this country from the taxes which bear on the productive classes and on the consumption of the bulk of the community, would be equivalent to making this nation tenfold greater than ever she has been.

I have advocated a property tax, not because I am

in favour of it, but because I see no other means of getting at present out of our difficulties; if a property tax were laid on we might then be enabled to reduce, not annihilate, the taxes which bear on articles which enter the consumption of the bulk of the people; as an illustration I take the article of sugar, on which nearly 5,000,000 $l.$ is at present raised, and I find that the consumption of sugar is but five ounces per week for each individual in the United Kingdom; now by reducing the tax on that article alone, and at the same time extending the market of supply, which has I apprehend been always forgotten in reducing duties, 14,000,000 $l.$ would be more easily collected on the article than 4,000,000 $l.$ is at present: therefore do I hold that the imposition of a property tax would probably be only for temporary purposes, until the elasticity of the country had enabled the population, intelligent, enlightened and industrious as they are, to recover their natural order of things.

That 10,000,000 $l.$ of taxes might be immediately struck off from the finances of this country is to me perfectly clear; I would substitute for it a tax on the funds, a tax on 20,000,000 $l.$ invested in East India securities, and on Bank property, on every particle of dormant capital; and instead of checking machinery, or checking freedom of trade, I would then give every possible stimulus to it, believing as I do that extended commerce removes want. . . . I would make property amenable in a greater degree to support the taxation of the country than it at present does, and by that means relieve labour from those burthens which render the employment of it less productive to farmers, manufacturers, and merchants for the security of property itself: I hold that property in this country has become daily more and more insecure from the distressed state of the people, and that finally (may it be averted) physical power will be brought into operation. . . .

In adopting the principle of a direct tax, I would say it was an evil we cannot avoid; but I see no other objection to it, nor indeed is inquisitorialness a paramount objection, as an income tax exists in other parts of the empire, where no objection is made to its inquisitorial character. Indeed the objections which have been so strongly urged against a tax on income on account of its inquisitorial character, ought not to outweigh the just consideration of equally apportioning the burthens of a country so heavily taxed as this, according to the means of those who have to pay them. . . .

I believe that the excessive taxation levied on articles of prime necessity consumed by the labouring classes, is the main cause of the distressed condition of the labouring classes in this country. . . . I think that the physical and moral deterioration now taking place among the labouring population of this country is owing mainly to excessive taxation pressing on the articles required for the daily support of life; and superadded to this cause is our imperfect monetary system, depriving the body politic of a sufficient circulating medium. Removing 10, 12 or 15 millions of taxes from articles of the greatest necessity, entering into the consumption of the labouring and middle classes, and levying the amount so removed in an equitable manner upon the property and income of the wealthier classes, in a fair and just proportion, would have the effect of materially improving not only the condition of the labouring classes, but of all other classes in the community. I believe it would renovate the social fabric; I believe it would cause the diffusion of our commerce to an immense extent; we should be able to compete with foreigners in their own market, instead of being excluded, as we now are becoming, from most of the channels of our foreign commerce: the French, the Germans, and even the Russians, are competing with us in markets of which we had conceived we had a monopoly; they are underselling us in cottons, in woollens, and in hardware; and if we go on with the present system, we shall lose the greater part, if not the total, of our foreign commerce, and be undersold even in our own colonies. As regards the higher classes, whatever affects the base of the pyramid, the million, intimately affects the condition of the higher classes, and I hold that the condition of the higher classes of this country, and their prosperity and happiness, is intimately interwoven with that of the poorest labourer in the country, and I really believe the higher classes in this country feel an intense interest in the welfare of the lower classes. The plan I propose would not finally take from the higher classes any proportion of their incomes; a small proportion abstracted at present would be returned with tenfold interest. One of its effects would be to increase the security and thereby improve the value of property, which is in my opinion rendered insecure by reason of the discontent, dissatisfaction and distress of the bulk of the community. It would have most materially the effect of removing a considerable portion of the excitement and dissatisfaction prevailing in this country so generally among the labouring classes. It would give to England a healthy tone of society; it would give her a better, a more moral, a more religious population; and it would raise her higher and higher than she has ever been yet among the kingdoms of the earth; such is my firm conviction. . . . A more just and equal distribution of the public burthens amongst all classes of persons in this country, levied upon the principle of each of these classes contributing as far as may be in proportion to their means, rather than by the present system of taxation levied upon the necessaries of life, is one, if not the principal, of those means by which the Legislature ought to seek the improvement of the labouring classes. The Legislature ought immediately to relieve the working classes by the proposed income and property tax; I believe that to be one of the chief means by which relief may be afforded to this country. I look on boards of trade and other things as temporary expedients, as useful so far as a plank or spar is to a drowning man.

[Arguments against the repeal of the Corn Laws:]

Connected with the foregoing evidence is the question of the corn laws.

Mr. Longson, in reference to the query: Do you consider that the tax upon foreign corn, and all the other taxes that affect you, are the real causes of your sufferings; or do you still, after having had them called to mind, ascribe the sufferings of the weaver to that unprincipled reduction that prevails so much in the manufacturing districts? Answered, I must state that with great positiveness and earnestness, that, all other circumstances remaining the same, suppose there was no tax, suppose there was no corn law, I do not conceive that the condition of the weaver would be the least remedied at all. It would be said by those who have not been improperly termed grinders, "We will offer you less." They will give us less and less, till the weavers and other classes would not have a potatoe a year more by the alteration. It would be taken up by this heartless competition, and the advantage thrown away upon foreigners upon the Continent; and as regards those manufacturing gentlemen of Manchester and those places, who contend that there is no other mode of relieving the distress of the weaver whatever, except by a free trade in corn, I am quite confident it could not give the relief, and I think it may be demonstrated.

Philip Halliwell says: With respect to the present system of the corn laws, I happen to differ very materially from the prevailing opinions upon these subjects. . . . It is the greatest madness in the world to neglect the home trade in order to cultivate a foreign trade, which produces no profit. And as respects the agriculturists I think it is the soundest policy for a State to make itself independent of foreigners, particularly for the article of bread. Whether it be from the county-rates and the tithes, and the taxes, rather than from any other cause, that the price of bread is so much dearer in this country, than it is to foreign weavers, I always consider, as far as the tithe is considered, that every gentleman or nobleman that holds his land, holds it subject to that incumbrance, and that to repeal the tithes will not be any real advantage to the people; that in all cases where the tithe is redeemed, the land naturally rents higher; and suppose there was a general system of repealing the tithe, the benefit of that system, except the corn laws were repealed at the same time, would naturally fall into the hands of the land-owner, and would not benefit the operative in any way whatever. There ought to be a protecting duty laid upon all corn imported. I think the agriculturist ought to be protected: the land-owner, the farmer and the labourer; and I think also the attention of the Government should be drawn to protect me as a manufacturer. I found, upon the protection given to all those kinds of property, my claim that my labour should have the same protection given to it.

Richard Needham remarks that if the corn laws were to be repealed, unless some such measure as have been applied for in petitions be adopted, the hand-loom weavers of Bolton would not benefit by the repeal. No, not by one farthing, excepting a law was to take place that was to secure the present wages we have. If they were to repeal all the taxes in the country, it would not do us one farthing of good: not that I am a friend to taxation; I am a friend to economy and moderation in all things; but if we could live upon a shilling a day, the system would take care we did not get 9d. to live upon; that is my idea, and all those I associate with. The circumstance of 280,000 weavers and all their households enjoying a competence to purchase the necessaries of life and the comforts of life would better the situation of the landed interest; we should consume more clothing; we should consume more butcher's meat and more corn of all descriptions, and that would cause a reaction favourable to the agricultural interests of the country, and it would react again upon us in the same ratio; it would be a common blessing.

John Lennon says: Supposing that by an alteration of the corn laws we could buy our food one-third cheaper, but that the present system of competition should go on, the weaver would derive very little benefit from the reduction of the price of food. I do not think we can derive any benefit unless we were protected by a minimum rate of wages, or by some system of uniformity of prices; a reduction of provisions would be very little interest to us; we could purchase very little, for the trade would be still declining.

John Scott asserts: If you reduce taxation, I have no doubt but there will be a very great benefit accrue to the working classes; but then I think that the state of trade has come to such a pitch that the manufacturer would do all he could to turn the profit into his own pocket. If we had no power-looms, if the manufacturer could not produce the articles which he would send to the foreigner by power, then indeed it would increase the demand for labour: but take the corn laws off, and the power-loom manufacturer will do all he can to increase his machinery to that extent that those goods may be manufactured at the lowest possible price at which they could be obtained. The manufacturer would exchange his cheap goods, upon which no manual labour was performed, with the foreigner, and would receive corn in return; and thereby he would become a twofold monopolist.

With reference to the question, whether it would be any advantage to the British weavers to gain the inhabitants of Poland as customers, if they lost as customers the labourers of the farmers of England?—I conceive that home consumption ought to be looked at first; for if the agricultural labourers of the kingdom were well clad with our manufactures, it would turn to our profit more than if we supplied the Poles. The agricultural labourers and the farmers of England cannot be well clad and use the goods that we weave, unless they have a remunerating price for their labour and their produce.

With reference to the abolition of the corn laws, it would be most unjust to a man who had invested his property in land under the faith of these restrictions, to abolish them without taking off those burdens which press upon him; and I am not one of those who suppose that a great and most valuable interest of this country, the landed interest, should be pulled down for the sake of some mere chimera, which it could not but be, if the present system of taxation were continued.

Report on Child Labor

This report on child labor was largely the result of the work of Sir Robert Peel (father of the prime minister). He was influenced by Robert Owen to advocate for limiting child labor by law. In 1802 Peel worked to get Parliament to pass the 1802 Health and Morals of Apprentices Act, which concerned poor children who had no parents to care for them and lived in workhouses. The workhouse governors apprenticed these children to work in factories for a certain number of years. The 1802 act made it illegal to employ these paupers in cotton mills for more than twelve hours a day. In succeeding years Peel worked to extend this protection to all children employed in cotton mills, and to make it illegal to hire children under the age of ten. On 3 April 1816 Peel proposed to the House of Commons that a committee study the issue. The following extract is a report of the evidence the committee gathered. This investigation resulted in the 1819 Cotton Mills and Factory Act, which made it illegal to employ children under nine years of age; children under the age of sixteen could not work more than twelve hours a day. This act applied only to cotton factories and did not have a strong mechanism to enforce it, as can be seen in the 1832 Sadler Commission report.

(For debates in Parliament on the 1818 bill, see T. C. Hansard, The Parliamentary Debates from the Year 1803 to the Present Time, *33:169–75, 342–74, 548–49, 578–82, 646–49, 793–96, 884–87; 38:169–75, 342–72, 548–49, 578–82, 646–49, 792–96, 1252–53; 39:287–89, 339–48, 652–56.)*

The question of child labor was hotly debated throughout England during the early nineteenth century. Consider carefully the competing claims about the health impacts of the factory system on the young children who worked in the factories.

There are two other questions that you should carefully consider as you prepare to debate child-labor laws: the efficiency costs of limiting child labor in factories, and the preferences of families in Manchester at this time.

- *Consider how competing claims about liberty and independence are likely to be framed regarding child labor and the family. Proponents of child labor will note that the wages children provide are often crucial to the family budget, while opponents of child labor will decry the effect this has on the family's moral development. What would your character recommend as a remedy—a wage for adult laborers that eliminates a family's need for child labor? Or continuing a practice that is currently supporting many poor families? How will you respond to arguments against your position?*
- *According to proponents of child labor, how extensive would the costs be to manufacturers if they eliminated child labor, or limited the working hours of children? How would your character respond?*

Source: "Report of the Minutes of Evidence, Taken before the Select Committee on the State of the Children Employed in the Manufactories of the United Kingdom, 25 April–18 June 1816," *Parliamentary Papers*, vol. 3.

Sir Robert Peel, Baronet, in the Chair.
Mr. Archibald Buchanan [*a supporter of child labor*], **called in, and Examined.**

What is your employment?—I am employed in the management of the cotton mills in Scotland. . . .

What number of persons are employed in their different works?— . . . there are 875 employed at the Catrine works.

How-many of those are under ten years of age?—Twenty-two males and thirty-seven females.

What is the youngest labourer that you employ?—I cannot answer that question; I suppose the youngest may be eight or nine: we have no wish to employ them under ten years of age.

Wishing not to employ any under ten, what circumstances have led you to employ any under that age?—The circumstances, generally, of the condition of their parents; people with large families, who find great relief from having a child or two put in at an earlier age. . . .

What are your hours of work?—Our working hours are twelve hours in the day.

At what time do they begin in the morning?—They begin at six o'clock in the morning, they stop at half-past seven at night, and they are allowed half an hour to breakfast, and an hour to dinner....

What has been the state of the health of those children, particularly those under ten years of age?—Generally very good; much the same as those children in the neighbourhood who are not employed in work.

You have not observed that the twelve hours work has interfered with the health of the children?—I have not.

The work requires more the attention of the eye and the hand, than labour?—The work is little or nothing with the young children, they have merely to attend there.

Suppose that the children were taken at six years of age, do you think they would be able to work that number of hours, without great indisposition?—I should think they would.

That it would not injure their health?—I have seen many instances of children that were taken in even as young as six, whose health did not appear at all to suffer; on the contrary, when they got to greater maturity, they appeared as healthy stout people as any in the country.

Not crippled in their growth?—No....

Do you ever receive complaints, from the parents, of the number of hours the children work?—I do not recollect any complaint on that subject.

Are the parents generally very desirous to send their children to you, or not?—Very desirous....

What are the weekly wages that a child of nine years old will acquire in your works?—All our children are paid by the produce of the rooms in which they work, or by the produce of the spindles which they actually attend.

You may probably be able to tell the lowest rate and the highest rate?—The children of nine years are generally learners, and receive 6 d. to 2s. per week, according to their ability....

Do you find that where manual labour is used, the wages fluctuate more than in your manufactory?—The hand-weaving fluctuates very much....

Is your employment more uniform than in other trades where there is less machinery?—Of course.

Then do you conceive that greater uniformity, both as to wages and employment, has a happy effect upon the wages of the people?—Very much so.

What would be the consequence of any interruption of your machinery, during the hours you are now working, supposing it to be stopped for half an hour prior to your stopping for dinner?—If we were allowed to work half an hour longer, to make that up, I cannot see any difference.

Would not the stopping of your engines be attended with an additional expense to you?—No doubt it would; if we stopped the tire engine or water wheels, we must stop the machines, and produce so much less.

If it was required of you to stop and dismiss those children half an hour prior to the time you now do, would not that produce a great inconvenience to you?—It would certainly enhance the cost of our produce materially....

Do you conceive that if by Act of Parliament the hours of labour were reduced, the poor for the reduced hours would have the same wages paid them they have for the greater number of hours—We could not afford the same wages, in the present state of the market....

Would not parents and other persons in your works consider the lessening of hours a considerable disadvantage to them?—No doubt.

Do you conceive you could afford, or would it be just, that the same wages should be paid for less work?—We could not afford it in the present state of the market, and we should not only be obliged to reduce the wages in proportion to the time lost, but we should have to reduce further, to make up for the deficiency of produce from our machinery, which affects our sunk capital....

Then would not any interference on the part of the Legislature be likely to be injurious to the trade?—I should think it would....

Are not the works of the grown-up people and

those of the children so intimately connected, that if you shorten the hours of the children's working, you must also shorten the hours of the grown-up people working?—Yes.

And consequently the machinery must be at a stand?—Yes. . . .

Have you known any serious inconvenience arise from the flue of the Cotton getting into their eyes?—I never knew any bad consequences arising from it. . . .

Some part of your works are employed in weaving by power?—Yes.

Do you employ generally men, or women?—Women, or, more properly speaking, girls from twelve to sixteen or eighteen, and probably twenty years of age.

What is the nature of their employment?—The nature of the employment is to attend to the loom, and mend the threads when they break down.

Is there any labour?—No, there is no labour.

Watching, you call it?—Yes.

Your competitor in the market, for the sale of goods so manufactured, is goods made by the hand-weaver?—Yes.

Do you happen to know the hours of labour generally of hand labourers, either in England or Scotland?—I do not; they work very irregularly.

Is that hard labour?—The hand-weaving is certainly pretty hard labour, where it is coarse goods.

Do you know whether children are employed in Lancashire in hand-weaving?—I cannot say to Lancashire; I have seen children working in Lancashire and other parts of England.

The body is put into an uneasy posture?—Yes, they sit leaning over the beam that is before them.

Have you had opportunities of seeing the shops in which those persons work?—I have frequently been in them.

Compared with your own establishment, what degree of comfort do those people enjoy?—I should think their workshops are much less comfortable.

Are they generally damp?—All hand-weaving is done in damp shops.

Have you known weavers employed very long hours; have you generally understood in the country, that they work fifteen and sixteen hours, at particular times?—I have heard some of the weavers say so. . . .

Is it very difficult to procure work?—I understand it is. . . .

Is not the hand-weaving done in the posture of sitting?—It is.

Do not you conceive that to little children, standing thirteen hours will be more injurious to health than if sitting down part of that time?—They do not stand in one posture, they have to move about the machine, which performs the motions requisite itself; and they occasionally sit, they have seats. . . .

Does it not frequently happen in the spinning machinery, that there are intervals, particularly after the spindles have been recently supplied, in which the children have nothing to do?—In water-spinning they have less to do.

And may occasionally sit down?—Yes.

Children are not much employed in mule-spinning?—A good many of them are employed as piecers, to mend the threads.

Of what age?—All ages almost; I do not know of any employed under, probably, seven or eight.

They are not employed to superintend the mule itself?—No, merely to mend the threads.

Do you find the children employed at your works, are in all respects as comfortable, and as well informed, as the children of parents in similar situations in life are?—I conceive they are generally better.

Have many of the children employed at your works gone to other employments afterwards?—The boys go to other trades, when they are grown up.

And have you generally observed that they are attentive to their masters?—I have heard their masters often say, that they preferred apprentices from the mills to those from the country.

From your observation, is the growth of the children affected by being in the factory?—Those who have grown up from the mills where I have been, appear to be as stout as any people in the country; they go to all trades, masons, and joiners, and weavers, and so on.

Are they as tall?—Yes, I do not see any difference.

Do you conceive that the habits of regularity they

are taught in the works, are advantageous to them in their pursuits afterwards?—I should think it was; that is the chief advantage tradesmen think they have in employing children from the factories.

That is, from their habits of industry?—Yes and the ingenuity they acquire in the works.

In what manner are children educated in your works?—We have three schools at present. . . . one of the schoolmasters . . . teaches one hour after the works stop on week days, and on Sundays he attends the Sunday schools; but I do not recollect the time the children attend.

You make the children's confinement fourteen hours, then?—They are not compelled to go to school in the evening. . . .

The education is not compulsory?—No.

But they have the means, if they choose to resort to them?—Yes. . . .

From what you know of the spinning of cotton, do you think the profits of that business have in general been great, during the last seven years?—From the number of failures which have lately taken place, I should think they were not.

Have the profits been equal to a fair return for the capital employed and the risks incurred in the business?—I should think not.

Have there been many bankruptcies among persons engaged in spinning and power-weaving of cotton, in Scotland, of late years?—This last year in particular there have been a great many.

Do you know whether many of your goods are sold on the Continent of Europe?—They are.

If your hours of labour were abridged by law, and the, cost of your goods thereby increased, as you have already stated, do you imagine any serious difficulty would be thereby occasioned to your house in the disposal of your manufactures?—No doubt there would.

Abroad?—Yes.

You therefore imagine that parliamentary interference, in the manner proposed, would increase your difficulties in regard to foreign competition, that is, competition in foreign markets with the produce of foreign manufactures of a similar description?—Yes. . . .

Matthew Baillie, M.D. [*an opponent of child labor*], **called in, and Examined.**

You are by profession a physician?—I am. Matthew Baillie,

You have had much experience in your profession?—I have. M.D.

Has your attention been directed to children?—Not perhaps so much to children as to grown people. . . .

At what age may children, without endangering their health, be admitted into factories, to be regularly employed thirteen hours a day, allowing them one hour and a half to go and return from meals, and one hour for instruction?—I should say, that there was no age, no time of life whatever, where that kind of labour could be made compatible, in most constitutions, with the full maintenance of health.

Do you think that children from seven to ten years of age could be employed more than ten hours per day, without injury to their constitution?—I think not; and if it was left to me to determine, I should say, that they ought to be employed fewer hours, for the full maintenance of health.

What do you consider to be the effect, upon the development and growth of the bodies of children from six to ten years of age, of so many hours confinement per day?—I cannot say much from experience, not having attended children that have been labouring in manufactories; but I can say, what appears to me to be likely to arise out of so much labour, from general principles of the animal economy. I should say, in the first place, that the growth of those children would be stunted; that they would not arrive so rapidly at their full growth; that they would not have the same degree of general strength; that it is probable their digestion would not be so vigorous as in children who are more in the open air and less confined to labour; that they would probably be more liable to glandular swellings, than children who are bred differently: and I think it likewise probable, that in particular manufactories at least, they would be more likely to be affected with diseases of the lungs.

What would you consider to be the effect on the mental faculties of children so young, when in addition to the facts already stated, the attention is constantly fixed and employed on one set of objects for days, months and years?—I should think, that they would acquire more acuteness with regard to that particular employment, but that with regard to all general employments and all general exercises of the mind, they would be inferior to what they would be, if their minds were directed to a greater number of objects either of curiosity or of study.

Have you considered at what age children might be safely employed in factories?—I can only answer this question by a kind of conjecture, which is found upon my acquaintance with the animal economy; I should say, that seven years old was perhaps the earliest age at which children should be so employed.

How many hours would you recommend children of that tender age to be employed?—I should say that at that age, probably, for the first year, they should not be employed more than four or five hours a day; and for the two succeeding years, they might be employed six or seven hours a day; that afterwards they might be employed ten hours a day, and beyond that, in my opinion, there ought to be no increase of labour. . . .

James Pattison, Esq. [*a supporter of child labor*], **called in, and Examined.**

. . . You are a silk manufacturer?—I am. . . .

How many persons are employed in those mills?—We employ from 320 to 330 persons.

How many of those are under ten years of age?—There are at present twelve from six to seven years of age, twenty-three from seven to eight years of age, and nineteen from eight to ten years of age, making in the whole fifty-four that are under ten years of age; those children earn from one shilling and sixpence to four shillings a week each.

How many are there in your works from ten to eighteen years of age?—A hundred and twenty-nine; and they earn from three shillings to eight and sixpence per week.

How many adults?—114 or 111; and they earn from nine shillings to fifteen shillings the men, and the women from four and sixpence to nine shillings a week. . . .

Why do you take the children so young?—The motive of taking the children so young is partly to oblige their parents; in a great degree to relieve the township; and also, because at that early age their fingers are more supple, and they are more easily led into the habit of performing the duties of their situation. . . .

What is the nature of the employment of the children under ten years of age; is it sedentary?—A child of that age is employed in looking after a dozen or twenty threads; the child has a place to walk up and down as long as this room is wide, nearly; and from the nature of the employment, the child never sits, excepting when his threads are all tied up which, when the silk is good happens frequently, so that he gets a good many resting times in the course of the day. . . .

Do you conceive that working in the factories is favourable to the morals of young people?—So far favourable to it, if I may venture to say so, that it keeps them out of mischief; and while they are industriously employed, they are less likely to contract evil habits than if they are idling their time away.

If a comparison were made, not between those who work in the factories and idle people, but between those who work in the factories and those who are industrious in other lines, what would you say to such comparison?—I should be rather puzzled to answer it; but I should presume in favour of those who are employed in the manufactories; the motive of my answer is, that I conceive that better state of the children would arise from the order and discipline which is maintained in manufactories. . . .

At what age do you take children?—From six to seven and eight; we never refuse any body.

Why do you take them at so early an age?—They

more easily acquire the facility of handling silk, and it is an advantage to themselves and their parents to take them at those early ages.

If you did not take children at that age, how would they be employed, do you suppose?—In running about the country, and in all sorts of mischief; at least such is the case with the children that I do not employ in the neighbourhood.

What number of hours are your children employed? —Including the dinner hour and their play hour, in the summer time they come at six in the morning and go away at six in the evening; and in the winter they come at eight and go away at eight.

Do you vary the number of hours according to their ages?—No; in the different departments they work together; in what we call the winding they all work together, and in the throwing they work together.

Do you conceive the employment of a silk-mill, or any part of it, to be laborious?—Not in the least; it consists wholly in opening the skeins of silk, and tying up the knots whenever it breaks; that is the whole labour attending it.

Do you consider it as unhealthy?—Not in the least....

If the hours of work were abridged, would their earnings be lessened?—Yes; my children do not work by the day; that is, not the whole of them; but they are paid for what they earn; I have established certain rules, which enable them to earn more or less, according to their industry.

If they were not employed by you, would the poor rate be affected?—Most undoubtedly; some two or three years ago I was under the necessity of stopping the manufactory to put a new water-wheel in, and I was engaged six months in that way, and during that time it fell very heavy on the parish, so much so, that in consequence of applications made to me by the parish officers and the parents, I paid them half what they otherwise would have earned, during the time the manufactory stood still....

Can you speak to the general health of the children? —I can, with great pleasure; I have not a sick person; there is no deformity, and no disease, except that there are a great many scrofulous families in the neighbourhood, and the children are infected....

Mr. George Gould *[an opponent of child labor]*, **called in, and Examined.**

You come from Manchester?—Yes....

What is the general appearance of the children working in the factories, as to growth, health, dress, and cleanliness?—In judging of children frequenting mills, by their size I have often taken them to be some years younger than they are, and the same remark has been made to me by others—generally speaking, their appearance is delicate, and less healthy than that of children confined to fewer hours of labour in the day; many of them, the younger children especially, are very badly clothed, and dirty. Upon the authority of a very experienced medical gentleman I state, that scrofulous complaints prevail very much amongst them, which he attributes chiefly to debility arising from excess of labour and confinement; and two other medical gentlemen asserted, that those complaints prevailed most in debilitated habits—the youngest children, moreover, from inattention and carelessness, often become crippled by being entangled in the machinery.

Those of a more advanced age are not so liable?— They have more circumspection as they get older.

What is their general moral character, and what effects have their employment and association upon their morals and intellects?—Their general moral character is bad; there exists amongst the children a great want of religious knowledge; and as the parents so much depend upon the earnings of their children, they lose the proper parental authority and command over them, and indulge them in Sabbath-breaking, and various other improprieties....

In your knowledge, is not the food of the children better now than it was fifteen years ago?—I think not....

You have stated in the course of your evidence, that the morals of the children in factories was not so good as the morals of children not employed in factories; the Committee wish to know upon what facts you stated that opinion?—In the first place, the children employed in factories are worked so long, that they use the Sunday as a day of recreation, and do not go for instruction any where, taking them in the majority; in the next place, the children that frequent factories make almost the purse of the family, and by making the purse of the family they share in the ruling of and are in a great state of insubordination to their parents.

Is the Committee to understand that this is the result of your own observation, or only general reasoning?—It is observation and reasoning combined. I have another thing to add, their association with the vilest characters of adults almost contaminates them, so that they have a worse chance, a great deal, than children who have not such association.

Do you mean the association that takes place in the mills, or out of the mills?—In going to and from the mills.

You have already stated that you have not had an opportunity of seeing them in the mills?—No, it is going to and from them.

Then the Committee is to understand, that this immoral effect is produced not in the mills, but in the association in going to and from the mills?—Yes. . . . The nature, of their employment is such, that in going to and from the mills, perhaps one half or three quarters of a year is in the dark.

And in going to and from the mills in the dark, you, are of opinion that the children are corrupted by society with those with whom they have been employed in the mills?—They hear the expressions of the worst characters, and perhaps are witnesses to some of their actions.

Then is the Committee to understand, that if they quitted the mills during the light, those immoral effects would be guarded against?—It certainly would be less; because I think that there are no parents who have any regard for their children, but wish that they should be as little in the dark as possible, without observation.

Would not in Manchester, whether in the dark or in the light, such immoral practices as you allude to, be observed and corrected?—It must be known to the Committee, that there is a common habit of swearing prevails, not only in Manchester but every where, I do not see any body who attempts to correct it.

Is that common habit peculiar to the mills?—I do not know that it is.

Then the immorality peculiar to the mills is not that of swearing because that is a common practice? —I say, I see no attempt to correct swearing, though it is well known.

The question alludes to disorderly conduct?— There is much rudeness: in expression and action, which contaminates the minds of persons associating with them, and yet there is nobody makes it his business to have it corrected.

Is there not in Manchester an appearance of great order and tranquillity in the streets at night?—I think there is.

And this appearance, with very little interference on the part of the police officers?—Yes, very little. I admit that.

And yet there are supposed to be from fifteen to twenty thousand persons employed in these cotton manufactories who are some of them discharged in the dark?—In reply to that, I should say, those Children are so exhausted, at least those who are the subject of the present inquiry, by labour, that they get home to get rest; they have not time in the week days. But see them when they are more adults on a Sunday, and then observe their conduct. I can truly say that their conduct on a Sunday is such, that females as well as men, insult often their well-behaved superiors in the streets, and that to such a degree, that well-behaved discreet gentlemen, even if they meet the factory people, will, if possible, go on the other side of the street to avoid them.

How do you distinguish on a Sunday between those that have been employed in factories, and those that have not?—On Sundays they are better dressed, and better fed, perhaps, and they have had a holiday, and therefore do not want to go to bed so soon. . . .

Does not that habit of independence, generated by good wages, produce those manners very much in manufacturing towns?—Certainly. . . .

Do you conceive that the Children so employed,

if they were dismissed an hour or two earlier, would go to school?—Many would. . . .

When they are dismissed early on a Saturday, do they go to school?—They are not likely, that is the sixth day of their labour in the week. . . .

Then you think that though they are employed a smaller number of hours on Saturday, yet they are still so oppressed with labour, that they are not in a condition to go to school?—They are less willing, and it depends on their own will, for the parents have very little authority over the children who support them.

Are you aware that if the hours of employment of children in factories were limited, the hours of employment of adults would be limited in consequence?—Yes, I am aware of that.

Then supposing the persons employed in factories were dismissed two or three hours earlier than they now are every afternoon, do you think that the tranquillity and order of the town of Manchester would be improved or injured by such an alteration?—Perhaps there might be a little more tumult occasionally, but I think it is necessary for the health and comfort of the adults, as well as the children, and would render them more disposed to go to public worship on Sundays than they now do, and thereby improve their morals.

Then do you think that it would be for the advantage of the adults to be prevented from labouring generally according to their own inclinations?—I think it may.

Then, if it would be for the advantage of the adult, in cotton factories, that the Legislature should prevent him labouring each day more than a certain time, would not that be equally for the advantage of those who do not work in cotton mills, but are engaged in other trades entirely dependent on manual labour?—I think there are more debilitating effects arising to adults from working in a cotton factory so long as they generally do, than there are in almost any other trade.

Is the labour of adults in cotton mills anything like the labour of weavers?—There is a longer confinement.

Is the labour nearly as fatiguing?—Long confinement upon our legs is very fatiguing, though one does little more than stand or walk.

Is not the labour of a weaver, and the labour of many persons engaged in manual operations in one hour, more fatiguing than the labour of persons engaged in cotton spinning is in two?—It is possible it may be more fatiguing, but he can relieve himself; in a cotton mill he cannot.

Do you mean to say that the posture of persons employed in cotton factories is always the same, and that they cannot relieve themselves by sitting down?—I think in many cases they cannot.

That they cannot in cotton manufactories relieve themselves by occasionally sitting down?—I think, from what I am informed, that they do not; I believe in, general the children do not sit down at all.

Have they no means of sitting down?—I have not heard that they have. . . .

Theodore Price, Esq. [*an opponent of child labor*], **called in, and Examined,**

Are you a magistrate of the county of Warwick?—I am. . . .

Have you ever visited any Cotton mill?—Yes, one. . . .

Did you make any observation on the state of the air in that Cotton factory?—I felt it very close when I went in.

Were the windows closed or open?—They were all closed.

At what season of the year was it?—It was in May; about the 12th of May. I asked how it was that the windows were closed when the weather was so warm, and the person who showed me the mill replied, that he did not open the windows if there was a breeze, that it blew about what he called the flew, which I understood to be the downy parts which blow off the cotton.

Did you make any particular observation upon that flew, or did you make any inquiry with respect to it?—Yes, I did; I said I thought it must be very prejudicial, he said, "yes, Sir, we are obliged to take emetics very often"; it struck me the whole atmosphere must be as it were saturated with it.

Was it very minute in its particles?—So minute I did not see it, except where it had gravitated in quantities, about the greasy parts of the mill.

Did any of the children ever take any emetics?—I saw a child in the kitchen, and the person who was showing me round, asked how the child was, a woman gave some answer, and he then said, "has the child had an emetic?" I had before been told they did take emetics in consequence of inspiring these downy particles.

What was the general state and complexion of the children employed in those works?—I thought the children short in stature, and they had a hectic appearance, probably from constant work and the warmth of the room; they were going backwards and forwards, twisting the threads when they broke, and they were obliged to move quickly across a large room there was a moisture about their faces, and they appeared hectic: I observed that in nearly all of them."

Did you make any inquiry with respect to the hours of their relaxation?—Yes, I did; I asked what recreation the children were allowed, they being employed so long a time in the mill, from twelve to thirteen hours; I think the man mentioned they came in at six in the morning, and staid till six or seven at night, I cannot tell which, and then it was that I asked what relaxation they had, and he said, they had an hour to dinner; and I replied, "I suppose, half an hour for breakfast," he said, no, they ate their breakfast in the mill; I said, "why that seems extraordinary, why should not you suffer them to go out of the mill to get their breakfasts?" he said, it put the mill out of gear, and it took considerable time to put it into gear again. . . .

Kinder Wood, Esq. *[an opponent of child labor]*, **called in, and Examined.**

What is your profession?—I am a Surgeon.
Where do you live?—At Oldham.
Near Manchester?—Yes. . . .

From your knowledge of the employment of children in Cotton manufactories, do you think that they might be employed, from 7 to 10 years of age, from 12 to 15 or 16 hours in the day, without injury to their health?—I certainly do not think that they ought to be employed from 12 to 15 or 16 hours in the day.

Do you think that such employment would be positively injurious to their health?—I think so; I think it would be injurious to my own.

Do you think children are less or more capable than adults of bearing labour?—I think certainly children are less capable of bearing labour.

That is, from 12 to 16 or 17 hours a day?—Yes; I think 12 is sufficient for any child to work in the day.

If excessive employment be found injurious to the health of children, does that appear immediately or gradually?—Not immediately, certainly; there is an affection which takes place in children when they are first sent to the mills, arising from the heat of the room. When a child first goes into the mill, he is placed in a very hot temperature, and an affection takes place, which is a slight fever; most children continue working nevertheless. This is not a contagious affection, but arises simply from being brought into a room of too high a temperature, and is preceded by considerable perspirations whilst in the works. I have witnessed this on many occasions. The effect of this feverish affection, and of the preceding perspirations, is debility, which is shewn in the paleness of the skin, which they rarely recover. When I say this, I mean that they seldom regain colour afterwards, but the affection is moreover less violent under different circumstances; the constitution of the individual itself influences it considerably.

What effect has this employment upon the growth of the children?—I believe my opinions differ from those of many other persons; I believe that their growth is not at all stinted by this employment. The growth is not at all diminished, so far as regards height; but with respect to bulk and solidity, it is. I have examined a great many of these for the army, who have been brought up in these works; they are generally tall, and of a slender configuration. Seldom rejected upon account of a deficient height, even the young boys; I know this differs from the common opinion. I have many reasons for thinking so. . . .

What hour do they begin to work in the morning?—It is a rule to lose no daylight: they may be found working from 6 to 7, and from 7 to 8; the former in summer, the latter in winter.

What time then is allowed for meals?—I believe, indeed I know the common regulation to be an hour for dinner.

Is there any time allowed for breakfast; do they leave the mill?—The engine does not stop for it.

They eat and work then?—They do; they attend to the machinery, and eat at the same time; the same regulation is observed for the afternoon refreshment.

What is your professional opinion of this arrangement of the employment of children?—The dinner hour I think a very good one; but considering the question, not with a view to trade itself, I think the children ought to have a recess when taking any nourishment; when they work and eat at one and the same time, the food is not sufficiently masticated, and is not brought into the stomach in a proper state. I believe that haste in the performance of that operation, is very detrimental to the health. This is my professional opinion.

Do you know the temperature of Cotton mills in your neighbourhood?—Generally, at is from 78 to 80, but it depends greatly upon the fineness of the spinning steam pipes run through the room. . . .

What is your opinion of temperature in rooms, where the thermometer stands from 70 to 90?—That is very high indeed. . . . the vicissitudes to which children are exposed, is more extreme in winter than in summer, the temperature of the atmosphere at the freezing point being 32 degrees.

What is your opinion of the effects of confinement, upon the health of children in Cotton mills, with such a high temperature?—I think the temperature, as stated, is very prejudicial to their health. . . .

Did you make any observation as to the ventilation of the rooms which you have spoken of; how was that managed?—There is no general arrangement for ventilation, so far as I know. . . .

Have you found that the floating cotton has been prejudicial to the health of persons working there?—I believe it very strongly; and from this circumstance, that the carders frequently become asthmatic. I mean tire men who have been long employed in these carding rooms are so affected.

Does it affect the children in a similar way?—It does not affect persons of a young age. It is a disease long in taking place; the aggregate of the circumstances influence the health of these children more than any one particular point.

Have you found any particular effect from the floating cotton upon the children, which would require the application of medicine at the time?—From the floating filaments I believe there is no immediate injurious effect, disease is long before it takes place. Without doubt the dust is inhaled into the lungs. About some of the machinery, the scutchers I believe they are called, the dust is very considerable; and the persons employed in that particular department work with a handkerchief brought over their mouth and nose.

In those scutchers were children employed?—Children were employed of about 14 or 16 years old in that part of the machinery.

Has it fallen within your knowledge, that persons have been so affected with the dust from the machinery, as to be obliged to give it up and go away to make room for others?—I cannot speak positively as to that fact. I can speak as to the fact that the dust was injurious to the animal economy, and so injurious that I should get out of it, if I were placed in it, as soon as possible. . . .

Have you observed the greater dangers to the children, arising from the machinery in low rooms, than in high ones?—The accidents are more common; the shafts and cylinders are lower, and the children are more likely to be caught up by them; sufficient room is not left for them to be carried round the shaft or cylinder. This arises from the shape of the room and the size of the cylinders, and therefore the work people involved are crushed to death. I have seen it too often, when we have been able to do nothing for them, and they have died before we reached the place. It may be well to state here, that where there is sufficient space between the top of the room and the cylinder, children will be carried once or twice round, and be very little injured.

Have you seen that often?—Yes, I have seen these fatal accidents too often.

Are there other accidents, which arise from the children getting their hands into the machinery simply?—That is a very common thing....

You have mentioned the hour when the work begins; have you, as a professional man, perceived inconveniences arising to the children from rising so early?—My opinion is, that the children certainly rise too early for their health. I apprehend, that a child that has any distance to go, cannot get to his work except he rises an hour previous; if he works twelve hours, and there is an hour before going to bed, and an hour after, that is fourteen, and an hour for dinner is fifteen, supposing him only to work twelve hours....

Have you perceived any effect on the digestive organs, from employment in Cotton mills?—The digestive organs become debilitated with the other parts of the system; it is a function that is easily injured; a child may eat a sufficient quantity of food, and not turn it into nutrition. If there is not a digestive power in the stomach, it is no matter what food he takes, the food is not converted to nutrition, he does not get any flesh....

Is it found to affect their general health and their longevity, after they have arrived at manhood, as well as their configuration?—I do not know instances of extraordinary longevity in our neighbourhood; instances of ordinary longevity are not very common amongst this class of people.

Are the children who work in factories, generally well fed by their parents?—I do not doubt they are, particularly when trade is good; I believe they live better than their betters.

Are they well clothed?—They are when trade is moderately good; there are few of them who have not two suits, a working dress and a better dress, sufficiently good. I do not conceive patches to be an objection.

Do you conceive them to be better fed and clothed than the children of poor parents who do not work in factories?—I believe they are pretty generally well fed in our neighbourhood, when trade is good....

Have you ever known a time when trade has been bad in the factories?—I know one now.

Do you see the effect of bad times in factory children on such occasions?—In their dress I do certainly....

Have you ever observed children allowed to sit down in the mills?—I cannot speak to that.

Have you ever seen them sit down?—No, I have not.

Do you conceive, as a professional man, that such a constant employment is consistent with the health of the young females who are employed in the mills?—I think it is too incessant;. I think there should be a recess from it for a short time....

Do you not find scrofula and consumption extremely common in the manufacturers?—Scrofula is common indeed, and consumption extremely common....

Are those complaints more common in the children employed in Cotton factories, than in those employed in other works?—I believe they are both; consumption is often the effect of a scrofulous disposition, and of those inflammatory affections of the chest that are produced by the sudden changes of temperature, arising from coming from a warm mill into a cold temperature, which is often a change of forty degrees.

Has it ever happened to you to observe children, with the scrofula upon them, continuing to work for a considerable period, without any material interruption arising to them from that disorder?—A child will work with scrofulous glands the same number of hours, but the constitution will sustain so much the more injury; they will very frequently work with diseased glands, membranes and bones; children very often work the full hours that have enlarged glands in the neck; children very often work the hours when white swelling is commencing in the bones of the knee. White swelling we esteem to be a scrofulous affection, and it exists in the membranes surrounding the knee and in the bones; but before those bones ulcerate, the blood vessels which supply them are in a state of increased action. I conceive a bone then to be diseased, though it performs the functions as far as regards the motions of the joint.

Such children are not laid by in an hospital on account of that disorder?—No; they continue to work as long as the disease permits them, or as long as the parents will permit them; that depends upon their means, or their tenderness, or their care. . . .

Do they receive any instruction in the course of the week?—There are a great many Sunday schools established in the neighbourhood; but when they work twelve hours, it is not very compatible with their taking instruction in the week day.

Are they too much fatigued to attend an evening school?—Yes; I think any body that worked labours would be in no great condition to go and take instruction. . . .

Are cases of bastardy very common in Oldham?—They certainly are very common. . . .

Have you known very early instances of young women being pregnant?—Yes, that is a department I practise in, and therefore I must know it because it happens frequently. I believe Oldham is not singular there; I believe London is as bad to the utmost, as far as my observations extend. . . .

On Taxation
The two most controversial taxes in Britain in the period around 1817 were the Corn Laws and the income tax, although there were many other taxes (direct and indirect, such as excise and customs taxes) that increased at this time.

On the Corn Laws
These laws applied to what the British call "corn"— that is, grains that require grinding, such as wheat, rye, and malt. The laws were intended to protect the British grain market from the competition of cheap foreign grain. During the Napoleonic Wars trade between Britain and Continental Europe was severely restricted. With the end of the wars, and the return of free trade, the British feared that the market would be flooded with cheap foreign grain and British grain producers (landlords and farmers) would be forced to lower their prices in order to compete. The Corn Law (Importation Act 1815 55 Geo. 3 c. 26]) made it illegal to import foreign grain into Britain if the price of domestic grain fell below 80 shillings per quarter (1 quarter = 480 lbs.).

The principle argument against the Corn Laws is that the burden of the laws falls most heavily on the laboring classes. What is the evidence for this claim? If your character is inclined to oppose the Corn Laws, you might also take a close look at the second half of our reading from The Wealth of Nations, *where Smith formulates an argument that restrictive trade laws tend to harm the wealth-creating capacity of the nation as a whole. What is his argument? This is important to consider in light of Malthus's argument in favor of the Corn Laws.*

In his Essay on the Principle of Population, *above, we saw Malthus worry that the operation of a free market on wages would tend inevitably to keep a large laboring class near or below poverty. Here is argues in favor of restricting markets in corn out of concern for what we might call national food security. If the Corn Laws are repealed, what does Malthus think would be the potential negative consequences for the country as a whole? What does he think would be the actual negative consequences for the wealthy land-owning classes? Why does Malthus think that repealing the Corn Laws wouldn't really help the laboring classes? Would your character agree or disagree?*

Questions to consider for both opponents and supporters of the Corn Laws: Is Malthus being inconsistent; that is, supporting a free market when it benefits the upper classes, but opposing free markets when it hurts their interests? Would repealing the Corn Laws benefit or harm Manchester as whole?

People of Southampton Opposed to the Corn Laws

Source: "Resolution of the People of Southampton on the Corn Bill (1814)," in William Cobbett, "To the People of Southampton, on the Corn Bill," *Cobbett's Weekly Political Register* 25, no. 23 (4 June 1814): 705–27.

"RESOLUTION 1st—That for several years past the price of wheat and other grain has been excessively high throughout this kingdom, and that the *consequent distress* has been considerably felt by all classes of society; while the poorest classes have occasionally been sorely and severely tried with all the evils inseparable from dearth and indigence."

"RESOLUTION 2d.—That this Meeting had earnestly hoped, in behalf of themselves and their poorer fellow-subjects, who have in general borne the calamities of the times with most laudable and exemplary patience, *that the return of Peace would have alleviated the distress that has been so long experienced, and would have carried comfort and plenty into every part of his Majesty's dominions.*"

"RESOLUTION 3d.—That this Meeting are struck with great apprehension as to the effects which they conceive will inevitably follow from the enactment of a Bill which is now depending in the House of Commons, on the subject of the Corn Laws; which must at once sweep away all hope of a reduction in the price of the most necessary article of human subsistence: fearful lest the *disappointment of expectations long cherished*, during a most protracted and anxious contest with foreign powers, should excite at home, among the suffering classes of the community, *a spirit of discontent and dissatisfaction*, at a moment when it is most fervently to be wished that this kingdom should find rest from that tedious course of suspense and calamity in which *foreign ambition and tyranny have so long involved it.*" . . .

"7thly.—That since, for so many years, *the middling and lower classes of His Majesty's subjects have borne the burthen and pressure of the times*, in a manner that reflects the highest honour on their good sense, and just value of the blessings of good government and social order, they have *a right to expect* that, in the present state [of] things, the *opulent landholders* of this kingdom should be prepared to make *some sacrifices*; that, in consequence of the excessively high price of corn, hay, and butcher's meat, since *the commencement of the war*, the landholders of the United Kingdom, on the expiration of leases held under them, have from time to time raised their rents from one to two hundred per cent and in many instances still higher, while *rectors and lay-rectors* have also, with *better reason*, raised their tythes in like proportion; so that these classes have thus been in a great measure, if not wholly, *indemnified against the taxes and consequences of the war*: while gentlemen (not being landholders), men of slender fortunes, annuitents, tradesmen, and the *poor at large*, could have no indemnity nor relief whatever; but were obliged to bear the heavy burden of the government and parochial taxes, both for themselves and for those exonerated as aforesaid."

"8thly.—That a Petition, grounded on these Resolutions, be presented to the House of Commons, praying that they will by no means sanction a plan that must inevitably fix the rent of land, at a permanently extravagant rate, confirm the *load of parochial burdens* for the maintenance of the distressed poor, render the most necessary article of subsistence perpetually dear, *bar the bounties of Providence from the majority of His Majesty's subjects*, and hopelessly discover the pleasing *association of peace with plenty* and cheapness, that has so long been a *source of consideration* in the midst of *extensive calamity*."

Times (London) Editorial Opposed to the Corn Laws

Source: Philanthropos, "To the Editor of the Times," *Times* (London), 23 February 1815, 3.

SIR, — The aphorism of Hume quoted in your leading article of this day—that "manufactures gradually shift their places, leaving those countries which they have already enriched, and flying to others *whither they are allured by the cheapness of labour and provisions*,"—ought to be written up in the Senate House of every commercial country. We should not then see our government busied in *raising the price* of provisions, to serve the present interest of a particular class of the community. I say their *present* interest, because if by the measures now pursuing, the manufactures be driven out of the country, the great expenses of our government and the enormous interest of our public debt must fall on the *landed interest* exclusively. They cannot shift their corn-fields or pastures to other countries. No, they must remain in the gap, alone, unfriended, and until the last gasp of political solvency. In the meanwhile there is no class in society that must not be injured. To the *Stockholder* the measure is palpably unjust; for by thus artificially raising the price of all articles of expenditure, his *interest* is virtually reduced, and the security of his *principal* more than endangered. To our *manufacturers* it is ruinous, for by compelling them to pay more for the subsistence of their workmen, than is done by their rivals in other countries, a protecting duty is imposed in favour of the foreign manufacture. Their profits must be immediately and greatly reduced to meet these rivals in the market, while in progress, even our own capital, skill, and labour, will glide into more genial stations, until one half of our manufacturers turn emigrants, and the other half are ruined in an unavailing struggle to preserve their footing in England.

To the *public at large* the measure is in every way oppressive and injurious. What! after twenty years of sacrifices and privations, have we at length arrived at peace, only to be told that we shall not enjoy its blessings?—that in order to perpetuate the vast advantages which the war has yielded to the landholder and the farmer, it is expedient that the *war prices* of the necessaries of life be continued?—that other property may fluctuate and sink to nothingness, without protection or pity, but that the gains of the landowner and his tenant may only *rise*—they must not *fall*? Surely this measure can never be carried, if Englishmen be true to themselves. You, Sir, have done much in exciting that patriotic spirit which ended in the downfall of the tyrant of Europe. I rejoice to find that your pen is directed against this alarming experiment on the vitality of England. We have seen the effect of a decided declaration of public opinion on the Property tax. The present question is one of far deeper and more lasting importance to the realm. Be the voice of the country declared with similar energy, and the permanent interests of England, and English industry, will not be sacrificed to the overweening desires of a particular class. I am, &c.

Thomas Malthus for the Corn Laws

Source: T. R. Malthus, *The Grounds of an Opinion on the Policy of Restricting the Importation of Foreign Corn* (London: John Murray, 1815).

[Malthus admits from the start that he is] . . . in favour of some restrictions on the importation of foreign corn. . . .

According to the general principles of supply and demand, a considerable fall in the price of corn could not take place, without throwing much good land out of cultivation, and effectually preventing, for a considerable time, all farther improvements in agriculture, which have for their object an increase of produce. . . .

It is impossible not to be convinced . . . that, during the last twenty years, and particularly during the last seven, there has been a great increase of capital laid out upon the land, and a great consequent extension of cultivation and improvement; that the system of spirited improvement and high farming, as it is technically called, has been principally encouraged by the progressive rise of prices owing in a considerable degree, to the difficulties thrown in the way of the importation of foreign corn by the war; that the rapid accumulation of capital on the land, which it had occasioned, had so increased our home-growth of corn, that, notwithstanding a great increase of population, we had become much less dependent upon foreign supplies for our support; and that the land . . . would be competent to the full supply of a greatly increased population: but that the fall of prices, which had lately taken place, and the alarm of a still further fall, from continued importation, had not only checked all progress of improvement, but had already occasioned a considerable loss of agricultural advances; and that a continuation of low prices would, in spite of a diminution of rents, unquestionably destroy a great mass of farming capital all over the country, and essentially diminish its cultivation and produce.

It has been sometimes said, that the losses at present sustained by farmers are merely the natural and necessary consequences of overtrading, and that they must bear them as all other merchants do, who have entered into unsuccessful speculations. But surely the question is not, or at least ought not to be, about the losses and profits of farmers, and the present condition of landholders compared with the past. . . . the real question respects the great loss of national wealth, attributed to a change in the spirit of our legislative enactments relating to the admission of foreign corn. . . .

I doubt whether in the most extensive mercantile distress that ever took in this country, there was ever one-fourth of the property, or one-tenth of the number of individuals concerned, when compared with the effects of the present rapid fall of raw produce, combined with the very scanty crop of last year!

Individual losses of course become national, according as they affect a greater mass of the national capital, and a greater number of individuals; and I think it must be allowed further, that no loss, in proportion to its amount, effects the interest of the nation so deeply, and vitally, and is so difficult to recover, as the loss of agricultural capital and produce. . . .

[It is] really impossible for us to grow at home a sufficiency for our own consumption, without keeping up the price of corn considerably above the average of the rest of Europe. . . .

It has, perhaps, not been sufficiently attended to in general, when the advantages of a free trade in corn have been discussed, that the jealousies and fears of nations, respecting their means of subsistence, will very rarely allow of a free egress of corn, when it is in any degree scarce. . . .

[If there was free trade and no Corn Law, Britain would import corn from France because they do not have enough for the subsistence of all the British people. But if there was a period of even moderate scarcity, the French would refuse to sell the British enough corn, because they would keep it for their own citizens. The British would be forced to look elsewhere for corn, and they might not be able to get enough such large quantities quickly enough.]

Such a species of commerce in grain shakes the foundations, and alters entirely the *data* on which the general principles of free trade are established. For what do these principles say? They say, and say most justly, that if every nation were to devote itself particularly to those kinds of industry and produce, to which its soil, climate, situation, capital, and skill, were best suited; and were then freely to exchange these products with each other, it would be the most certain and efficacious mode, not only of advancing the wealth and prosperity of the whole body of the commercial republic with the quickest pace, but of giving to each individual nation of the body the full and perfect use of all its resources. . . .

There is, on this account, a grand difference between the freedom of the home trade in corn, and the freedom of the foreign trade. A government of tolerable vigour can make the home trade in corn really free. . . . But it is not in the power of any single nation to secure the freedom of the foreign trade in corn. To accomplish this, the concurrence of many others is necessary; and this concurrence, the fears and jealousies so universally prevalent about the means of subsistence, almost invariably prevent. There is hardly a nation in Europe which does not occasionally exercise the power of stopping entirely, or heavily taxing, its exports of grain, if prohibitions do not form part of its general code of laws.

The question then before us is evidently a special, not a general one. It is not a question between the advantages of a free trade, and a system of restrictions; but between a specific system of restrictions formed by ourselves for the purpose of rendering us, in average years, nearly independent of foreign supplies, and the specific system of restricted importations, which alone it is in our power to obtain under the existing laws of France, and in the actual state of the other countries of the continent.

In looking, in the first place, at the resources of the country, with a view to an independent supply for an increasing population; and comparing subsequently the advantages of the two systems above-mentioned, without overlooking their disadvantages, I have fully made up my mind as to the side on which the balance lies; and am decidedly of opinion, that a system of restrictions so calculated as to keep us, in average years, nearly independent of foreign supplies of corn, will more effectually conduce to the wealth and prosperity of the country, and of by far the greatest mass of the inhabitants, than the opening of our ports for the free admission of foreign corn, in the actual state of Europe....

1. And first let us look to the labouring classes of society, as the foundation on which the whole fabric rests; and, from their numbers, unquestionably of the greatest weight, in any estimate of national happiness.

If I were convinced, that to open our ports, would be permanently to improve the condition of the labouring classes of society, should consider the question as at once determined in favour of such a measure. But I own it appears to me . . . that it will be attended with effects very different from those of improvement. We are very apt to be deceived by names, and to be captivated with the idea of cheapness, without reflecting that the term is merely relative, and that it is very possible for a people to be miserably poor, and some of them starving, in a country where the money price of corn is very low. . . .

In considering the condition of the lower classes of society, we must consider only the real exchangeable value of labour; that is, its power of commanding the necessaries, conveniences, and luxuries of life. . . .

A high money price of corn would give the labourer a very great advantage in the purchase of the conveniences and luxuries of life. The effect of this high money price would not, of course, be so marked among the very poorest of the society, and those who had the largest families; because so very great a part of their earnings must be employed in absolute necessaries. But to all those above the very poorest, the advantage of wages resulting from a price of eighty shillings a quarter for wheat, compared with fifty or sixty, would in the purchase of tea, sugar, cotton, linens, soap, candles, and many other articles, be such as to make their condition decidedly superior.

Nothing could counterbalance this, but a much greater demand for labour; and such an increased demand, in consequence of the opening of our ports, is at best problematical.

The check to cultivation has been so sudden and decisive, as already to throw a great number of agricultural labourers out of employment; and in Ireland this effect has taken place to such a degree, as to threaten the most distressing, and even alarming, consequences. The farmers, in some districts, have entirely lost the little capital they possessed; and, unable to continue in their farms, have deserted them, and left their labourers without the means of employment. . . .

Our commerce and manufactures, therefore, must increase very considerably before they can restore the demand for labour already lost; and a moderate increase beyond this will scarcely make up for the disadvantage of a low money price of wages. . . .

On the labouring classes, therefore, the effects of opening our ports for the free importation of foreign corn, will be greatly to lower their wages, and to subject them to much greater fluctuations of price. And, in this state of things, it will require a much greater increase in the demand for labour, than there is any rational ground for expecting, to compensate to the labourer the advantages which he loses in the high money wages of labour, and the steadier and less fluctuating price of corn. . . .

[With the return of peace to Europe it is likely that the British will lose some markets for their manufactured goods on the Continent.]

Under these circumstances, it seems peculiarly advisable to maintain unimpaired, if possible, the home market, and not to lose the demand occasioned by so much of the rents of land, and of the profits and capital of farmers, as must necessarily be destroyed by the check to our home produce. . . .

Already, in all the country towns, this diminution of demand has been felt in a very great degree; and the surrounding farmers, who chiefly support them, are quite unable to make their accustomed purchases. . . .

3. Of the class of landholders, it may be truly said, that though they do not so actively contribute to the production of wealth, as either of the classes just noticed, there is no class in society whose interests are more nearly and intimately connected with the prosperity of the state. . . .

Of the effect, therefore, of opening the ports, [i.e., if there is free trade and there are no Corn Laws] in diminishing both the real and nominal rents of the landlords, there can be no doubt; and we must not imagine that the interest of a body of men, so circumstanced as the landlords, can materially suffer without affecting the interests of the state. . . .

[The value of the rents of landlords] . . . is not a mere benefit to a particular individual, or set of individuals, but affords the most steady home demand for the manufactures of the country, the most effective fund for its financial support, and the largest disposable force for its army and navy. . . .

[If the Corn Bill is not passed and the price of grain goes down, the nominal revenues of] the industrious classes of society and the landlords . . . compared with the average of the last five years, will be diminished one half; and out of this nominally reduced income, they will have to pay the same nominal amount of taxation.

The interest and charges of the national debt . . . are now little short of forty millions a year; and these forty millions, if we completely succeed in the reduction of the price of corn and labour, are to be paid in future from a revenue of about half the nominal value of the national income in 1813.

If we consider, with what an increased weight the taxes on tea, sugar, malt, leather, soap, candles, &c. &c. would in this case bear on the labouring classes of society, and what proportion of their incomes all the active, industrious middle orders of the state, as well as the higher orders, must pay in assessed taxes, and the various articles of the customs and excise, the pressure will appear to be absolutely intolerable. Nor would even the ad valorem taxes [i.e., property taxes] afford any real relief. The annual forty millions, must at all events be paid; and if some taxes fail, others must be imposed that will be more productive. . . .

From this review of the manner in which the different classes of society will be affected by the opening of our ports, I think it appears clearly, that very much the largest mass of the people, and particularly of the industrious orders of the state, will be more injured than benefited by the measure. . . .

I firmly believe that, in the actual state of Europe, and under the actual circumstances of our present situation, it is our wisest policy to grow our own average supply of corn; and, in so doing, I feel persuaded that the country has ample resources for a great and continued increase of population, of power, of wealth, and of happiness.

On the Income Tax

The income tax (a.k.a. property tax) was first imposed by William Pitt in 1799 and still in effect at the end of the Napoleonic Wars. It required that British subjects pay the government 10 percent of any income made from land, the commercial occupation of land, trading, public securities or dividends, professions, trades, offices, and employment. Subjects with income under £50 per year were excluded from this tax, and those with incomes between £50 and £150 paid a reduced tax. It was not renewed (thus abolished) on 18 March 1816 after a widespread petitioning campaign.

Two principle arguments are presented here against an income tax: first, that an income tax is tyrannical, and second, that imposing an income tax would not benefit the laboring classes. What are the reasons opponents of the income tax give for thinking that it would not benefit the laboring classes? How plausible would your character find these arguments, given that these taxes could be employed in furnishing public services for the laboring classes or for all of Manchester?

In the section on other taxes, we find an impressive list of the burdens that the laboring classes bear via so-called indirect taxes—taxes imposed on widely used goods such as beer, salt, and paper. How burdensome are these taxes? Could the laborer fairly regard the levels of taxes that they pay via indirect taxes as tyrannical?

In this section, note also the argument that when choosing among different forms of taxation we are making choice among relative evils. Why does the author think this? Which forms of taxation, from your character's perspective, would impose the least burdensome costs on Manchester as a whole? What is the fair share for your class and for other classes to pay? Why?

Times (London) Editorial Opposed to the Income Tax

Source: V., "To the Editor of the Times," *Times* (London), 15 February 1816, 3.

Sir—If the public press should now relax its exertions but for a single moment, the cause of the country is lost. The finance minister has opened to our view such a train of evils as were surely never before connected with the existence of a state of even nominal tranquility, and which, if the minister be suffered to realize them, will at once make the question of peace or war a matter of absolute indifference to the people.

The military peace establishment of England is fixed at 149,000 men. . . .

The war taxes have amounted to twenty-four millions per annum, originally submitted to by this liberal nation under the solemn pledge that they should expire with the war which had produced them. Of these, it appears that above sixteen millions are still to be levied on us, although the war is at an end, for the maintenance of that army whose amount is incompatible with the spirit of peace towards foreign countries, and with the preservation of liberty at home.

Under these painful circumstances, I cannot forbear to thank you for the lead which you have taken amongst the public writers of Great Britain in your early resistance to the income tax, the most unprincipled and hateful of all measure that could be devised for breaking down the spirit of a great people, and enabling their rulers to command, by an overwhelming armed force, the tyrannical exposure of any private transaction—the rude violation of the most sacred sanctuaries of honour, faith, and feeling amongst men. . . .

The income tax . . . is a lay inquisition: and we should object to it . . . for its fatal tendency to undermine the whole fabric of English freedom—and, in conjunction with the baneful but now incorrigible system of excise laws, to corrupt the very essence of our national character and happiness. . . . The despotic spirit of this inquisitorial impost, with its brood of petty tyrants in every village through the land, is the true and vital objection to our admitting it as a permanent branch of our financial system. A single root of arbitrary power will insensibly throw out shoots and suckers on all sides—so a government, exercising inquisitorial prerogatives in the collections of a single tax, will easily build upon this precedent of tyranny, and the subject, used to submit in one case, tolerates his degradation in many more. . . .

It is said, again, the property tax falls lightest on the lower orders of the people; and if we reject that mode of taxation, we must have recourse to some other of equal amount, and of greater and more unsparing severity. The proposition contains a double fallacy.

In the first place, the property tax does fall upon the poor, because it impoverishes the employers of the poor, and narrows the market for their labour. But the second member of the argument is the most extraordinary, illogical, and unblushing, that we have ever seen. *Why* is it . . . necessary that we should raise another tax . . . to supply the place of a war tax, proposed by minister for the maintenance of a peace establishment? . . . Where then, we ask, is the necessity for a peace establishment so enormous as to require the continuance of burthens justifiable only by a state of war? . . . this most ungrateful reward for [the English people's] . . . twenty-five years of efforts and of sufferings. . . .

Does any rational being suspect that we are now on the eve of a French, or of any other invasion? . . . . We have gone too far in our notions of military grandeur; the apparatus of armies, continued for a length of years, gives a tone to the government and people of a country rather different from that of our mixed constitution. It is in peace only that any effectual stand can be made against the growth of the monarchical and soldier spirit, or any ground recovered of what has been lost to popular and independent feeling. . . .

Times (London) Editorial against Taxes on Property

Source: *Times* (London), 19 January 1815, 2.

. . . The difference between the income tax and any other tax, is the difference between drawing blood from the veins, and drawing sweat from the brow,— between plucking off a branch, and tearing up the roots. In taxing articles of consumption, we do not touch capital, until after it has gone the full round of production: in taxing income, we *pro tanto* deprive capital of its productiveness. From that ill-fated moment when Mr. PITT, directly contradicting a principle he had maintained but the year before, proposed a Tax on Property, we have never been able to repeat the boast of BURKE that "with us, labour and frugality, the parents of riches, are spared and wisely spared." As to the pretended fairness of this oppressive tax, it never existed, and never could exist, but in theory. . . .

On Other Taxes

The income tax was in addition to a wide variety of other direct taxes (for example on windows and houses, on stamps for legal documents, and on luxury items like servants, carriages, horses, dogs, hats, and hair powder). There also were many indirect taxes, such as excise duties (i.e., inland taxes on goods sold domestically) covering many things, like beer, malt, spirits, wine, soap, salt, coal, glass, leather, tea, coffee, tobacco, and some raw materials. Finally, customs duties (i.e., border taxes on the importation of foreign goods) were imposed on over 1,000 imported goods. (On the direct taxes, see, for example, Kearsley's Tax Tables: Containing the Acts Relating to the Property Tax . . . [London, 1809].)

Edinburgh Review *Article on* Indirect Taxes and Poverty

Source: "Taxation and the Corn-Laws," *Edinburgh Review* 33, no. 65 (January 1820): 155–87.

. . . The industry of a great commercial country, is always liable to temporary embarrassments. . . . But we believe that Great Britain, since the return of peace, affords the only instance of a regorgement being simultaneously felt in every employment in which capital had been invested. The universality of the present distress forms its distinguishing and characteristic feature. . . . Pauperism, instead of being diminished, is rapidly increasing: Nor, without some very decided change in our domestic policy, is there the least reason to expect any material improvement in the condition of the great body of the people. . . .

As might have been expected, a variety of conflicting and contradictory statements have been made respecting the causes of this alarming increase of pauperism. . . . *[The author does not believe that the transition from war to peace, or the operation of the Poor Laws are primarily responsible for the poverty.]*

Other causes have unquestionably conspired to produce this effect; and of these, it will be found, that Taxation, and the restrictions on the trade in Corn, have been decidedly the most powerful.

In the present improved state of the science of political economy, it is unnecessary to set about proving that a heavy taxation on the principal necessaries of life, must be extremely prejudicial to the great body of the people—to all who either depend for subsistence on the wages of labour, or the profits of stock. This is admitted on all hands; but it has been strenuously denied, that these effects can be justly ascribed to the system of taxation adopted in this country: And as it is of the utmost importance, in every inquiry into the causes of the public distresses, that we should have correct opinions on this fundamental point, we shall avail ourselves of this opportunity to premise a few observations on the effects which must in general result from the imposition of heavy taxes on necessaries, before examining the nature and operation of the system of taxation to which we are now subjected.

In countries, such as the United States, where there is a boundless extent of fertile and unappropriated land, and where no feudal privileges or impolitic restraints fetter the employment of industry, or retard the accumulation of capital, the imposition of a tax on a commodity necessary for the subsistence of the labourer, would not be attended with any very injurious effects. In such countries, both the profits of stock and the real wages of labour are high; and a considerable revenue might be collected without occasioning any great inconvenience either to the workman or his employer: A little economy would enable the former to save the amount of the tax out of his wages; and these might be advanced without the rate of profit and the power to accumulate capital being thereby materially impaired. But in all old settled and fully peopled countries, taxation is infinitely more injurious. The supply of labour being in this case almost always greater than the demand, the real wages of labour are comparatively low; while, from the necessity of cultivating inferior soils, the profits of stock are also comparatively limited. In a country thus circumstanced, there is obviously very little room for increased economy; nor can a rise in the price of necessaries, that is, of those commodities "which the custom of the country renders it indecent for creditable people, even of the lowest order, to be without,"* be compensated by an immediate corresponding rise of wages.—The labourer is, in this respect, placed in a much more disadvantageous position than either the master manufacturer or capitalist.—When a tax is imposed on raw produce, or any species of manufactured commodities, the producers, by limiting the supply, are enabled to raise the price to such a sum as will afford them, exclusive of the tax, the common and ordinary rate of profit on their capital But this is a resource from which the labourer is in a great measure cut off. He is unable to raise his wages in proportion to the increased price of the commodities he consumes; and for this obvious reason, that, while the competition for employment, or the number of labourers continues undiminished, the demand for their services, however much it may be lessened, cannot be increased by the imposition of the tax.—The supply of workmen is not like the supply of boots and shoes; it does not and cannot be made to vary with every variation in the price of necessaries, or the rate of wages, whatever degree of stimulus may have been previously given to the principle of population, it is plain that, although the demand for labour should be suddenly contracted, or, which is the same thing in effect, though the proportion of wages to prices should be suddenly reduced, it would, notwithstanding, continue flowing into the market with nearly the same rapidity as before: Nor would the ratio of the increase of population be materially diminished, until the misery occasioned by the restricted demand on the one hand, and the increased supply on the other, had been very generally and widely diffused.

The principle, therefore, which has been laid down by Dr Smith, and other political economists, that every direct tax on wages, or on the commodities necessary to the subsistence of the labourer, falls entirely on his employer, must be received with very great modification: Except in the rare case where an unusual demand for labour occurs at the time that a tax is imposed on necessaries, it is impossible that wages should be equally raised. There is indeed but too much reason to believe that, in the great majority of cases, a very long period must elapse before any such effect can be produced. In the stationary state of society, or where capital and population are advancing with nearly equal degrees of rapidity, the more powerful operation of the principle of moral restraint, or a diminution of the rate at which population had previously increased, is the only way in which wages can be raised. But as this must be the work of time, there is an extreme risk lest the opinions and habits of the labouring class should in the interim undergo a change. When wages are diminished to any great extent, as they are sure to be by every considerable increase of taxation, the poor are obliged to economize; and it is natural to suppose, that what was at first forced on them by necessity, should ultimately become habitual. It is in this that the great evil of excessive taxation principally consists. Wherever the labouring classes are exposed to long-continued suffering and want, their opinions as to what is necessary for their comfortable subsistence, and the place they ought to hold in society, become degraded. The inadequacy of wages has already compelled the greater part of the people of Britain to relinquish a variety of comforts, and to satisfy themselves with comparatively coarse and scanty fare. And as the necessity for making still further retrenchments does not appear to be at all diminished, it is but too certain, if no means are taken to relieve the overloaded springs of industry, and to stimulate the natural demand for labour, that the ordinary rate of wages will be reduced to such a sum as will barely enable the labouring class to exist, and to continue their race. Whenever wages have been reduced thus low, it is true that they can sink no lower; and then, but not till then, the labourer will be beyond the reach of taxation; and every tax affecting the commodities indispensable for his support, will be paid by his employer, or, which is the same thing, will directly and immediately fall on the profits of stock.

It is impossible, however, to conceive a more wretched state of society, than that in which the bulk of the people are reduced to a dependence on mere necessaries, "In those countries," Mr Ricardo has well observed, "where the labouring classes have the fewest wants, and are contented with the cheapest food, the people are exposed to the greatest vicissitudes and miseries. They have no place of refuge from calamity; they cannot seek safety in a lower station; they are already so low, that they can fall no lower. On any deficiency of the chief articles of their subsistence, there are few substitutes of which they can avail themselves; and dearth to them is attended with almost all the evils of famine." Nor is this all:— Men placed in such circumstances, and cut off, as they must be, from all hope of rising in the world, naturally sink into a state of indolence and insensibility. They may not be discontented; but it is not in the nature of things that they should be either active or industrious. No man submits to privations and labour, but in the hope of obtaining corresponding comforts. Where there is no power, there can be no motive to accumulate; and, what perhaps is still worse, where the mass of the people are sunk in the abyss of poverty—where they have no *stake in the hedge*—it is impossible they should feel any great respect for the rights of those who have: And it is but too evident, that it is only by the terrors of the criminal law, that such persons can be prevented from breaking down those institutions which, however essential to the maintenance of society, must appear to them, not as bulwarks, raised for the public benefit, but for the support and protection of a favoured few. . . .

But, a direct tax on wages, or, which is the same thing, on the commodities indispensable for the support of the labouring classes, is not objectionable on the single ground of its having a constant tendency to degrade their condition in society; Taxation, in every form, presents only a choice of evils. Supposing, which is extremely improbable, that, notwithstanding the suffering and distress occasioned by the imposition of a heavy tax, *[the population increases at the current rate, and]* . . . wages are in the long run advanced proportionally to the tax; still the condition of society would be altered very much to the worse. The profits of stock would now be diminished in the precise proportion that wages had been increased. For, Mr Ricardo has demonstrated, that, whatever is added to wages, must be taken from profits; and conversely. Dr Smith, who was not aware of this fundamental principle, supposed that a heavy taxation on necessaries neither fell on the capitalists nor the labourers, but on the consumers generally; and that it was always in the power of the producers to indemnify themselves for a rise of wages, by enhancing the price of the commodities brought to market. But it is easy to see that no general rise of wages can have any such effect. Commodities are in every case bought by commodities; and as a rise of wages must affect, in an equal degree, the producers of every different article, it cannot possibly derange their relative values one with another, or occasion any increase of price.

It appears, therefore, that a slow and gradual increase of taxation . . . has a tendency to raise the rate of wages, and, consequently, to throw the burden from the shoulders of the labourer to those of his employer. But, even in this its least obnoxious shape, it is not easy to estimate all the evils it occasions. . . . The fall of profits consequent on a rise of wages . . . has a powerful effect in stimulating . . . *[the]* transfer *[of capital]* to other countries. The efflux of capital is one of the worst consequences of excessive taxation; and it is one against which it is impossible to guard. . . . The same principle which would prevent the employment of capital in Yorkshire, if it did not yield the same rate of profit that might be derived from investing it in Kent or Surrey, regulates its distribution among the different countries of the world. It is true, the difference in the rate of profit must be considerably greater, to occasion a transference of capital from one country to another, than from different provinces of the same country. But a comparatively heavy taxation is more than sufficient to occasion this difference. . . .

It is thus that heavy taxes on necessaries become, in the words of Dr Smith, "a curse equal to the barrenness of the soil, and the inclemency of the heavens." Such taxes must necessarily fall either on *wages* or on *profits*. To whatever extent they diminish wages, they must equally diminish the comforts and enjoyments of the largest and most important class in society, and spread pauperism, misery, and crime throughout the country; while, on the other hand, they cannot diminish profits, without occasioning a corresponding diminution of the power to accumulate capital, and without also stimulating its transfer to those countries, in which taxation is less oppressive. In the first case, their effect in degrading the condition of society, is instantaneously felt; in the second, it is brought about more slowly and circuitously; but in both, they are, in the end, nearly equally destructive of the happiness and future improvement of the society in which they have been carried to an inordinate extent.

But, if such be a tolerably correct estimate of the effects of heavy taxation on the condition of society, we can be at no loss to account for the increase of pauperism since 1793. During this period, the public burdens have been augmented to an extent unknown in any former age or country. No source of revenue, however trifling, and no necessary, however indispensable, not to comfort merely but existence, has been able to elude the grasp of the tax gatherer. Mr Pitt, and the subsequent Chancellors of the Exchequer, whatever may be thought of their merits in other respects, must be admitted to have had no equals in the devising of means to divert the greatest possible portion of the wealth of the country, into the coffers of Government. It is no exaggeration to affirm, that, with the solitary exception of water, there is not a single necessary consumed in the Empire, which is not, directly or indirectly, loaded with a most oppressive impost. Nor has the rapidity of the increase of taxation been less extraordinary, than the extent to which it has been carried. For example, the duty on tea, which, in 1793, was only 12 per cent, is now more than *eight* times as much, or 100 per cent. The duty on salt, which amounts (in England) to 15s. a bushel, or to about *thirty* times its natural cost, was *tripled* in 1805. The duty on leather, after being stationary for more than a century, was *doubled* in 1812. And the various duties on sugar, beer, spirits, soap, candles, tobacco, &c. besides the house-tax, window-tax, and stamp-duty, have all been increased in similar proportions. But, in order to show the progress of taxation, it is not necessary to engage in the endless and irksome task of enumerating the different articles on which new duties have been imposed, or the old ones increased. It is sufficient to mention, that the total payments into the Exchequer in 1793, on account of permanent and temporary duties, amounted to 17,674,395 *l.*; in 1804, they had increased to 49,335,978 *l.*, or to nearly three times their amount in 1793; in 1808, they exceeded the enormous sum of 66 millions; and in 1819, in the fifth year of the peace, they amounted to 47,990,814 *l.*, or to very nearly their amount in the eleventh year of the war. During the American war, the revenue, when greatest, never reached the sum of 13 millions!

Had this increased taxation sufficed to defray the entire expenses of the war, however oppressive in the mean time, its reduction on the cessation of hostilities would have enabled the country to avail itself of its many natural advantages, and again to spring forward in the career of improvement. This, however, was very far from being the case. . . .

Testimony to Parliament about Taxation
The following discussion of taxation is from a later period than the game. This means that the prices in 1817 would be different. But the argument made here was also common in the earlier period. Note: *The original has been altered to include paragraph breaks.*

Source: Great Britain, Select Committees on Hand-Loom Weavers' Petitions, 1834–1835, *Analysis of the Evidence Taken Before the Select Committees On Hand-Loom Weavers' Petitions (1834–1835): Ordered, by the House of Commons, to Be Printed, 10 August 1835* (London, 1835).

Section VII. Taxation.
. . . *Robert Montgomery Martin, esq.*, says, . . . I am the author of a work on the "Taxation of the British Empire," . . . I have directed my attention to the present social state of this country, and more particularly to the condition of the working classes, and the taxation which they pay in proportion to their wages. . . . I have prepared an analysis of a few of the articles bearing more particularly on the labouring classes, which I beg to deliver in:—

"Malt liquor I take to be one of the first; if a labouring man consume one pot of beer (porter) daily, the taxation, direct and indirect, on him is,—1st, On the land whereon the barley is grown; 2d, On the taxed labour which grows it; 3d, On the malt; 4th, On the maltster's charge for vexatious Excise regulations; 5th, On the hops; 6th, On the licence for a publican. On all these items, 3 d. out of the 4 d. is tax; therefore, on 365 pots of stout, the working man pays an annual tax of 4 *l.* 11 s. 3 d.

I come to the second article which enters into the consumption of the working man, Sugar; if a labouring man consume one pound of sugar weekly (which is allowed in some workhouses, and to the lowest household servant), the taxation, direct and indirect, thereon is, direct on the sugar 2½ d., indirect by West India monopoly, 1½ d., total 4 d.; therefore on 52 pounds of sugar the poor man pays an annual tax of 17 s. 4 d.

I come to Tea or Coffee, which enters into the consumption of the working classes; if a labouring man consume four ounces of tea per week (or its equivalent in coffee) valued at 4 s. per pound, the direct taxation thereon is full 100 per cent, therefore 12 pounds of tea per annum consumed imposes on him a direct tax (on a wholesome stimulant) of 1 *l.* 4 s. annually. If coffee be used instead of tea, it will not make any perceptible difference in the amount of the taxation, because the indirect mulct by reason of the West India monopoly of the coffee market fully compensates for the difference of tax levied; this is independent of the licence for permission to sell tea or coffee.

Soap; if a labouring man use one pound of soap weekly to wash himself and his clothes (I beg to observe I have been under the estimate in every instance), he pays a direct tax thereon of 2 d. per pound, and an indirect one of 1 d., more owing to harassing Excise laws and Custom duties on tallow, oil, barilla, &c. as well as on the taxed labour in preparing it; thus 3 d. per week for one year is 13 s. per annum.

On Housing; a labouring man it is to be supposed requires housing; for the poorest tenement, or part of a tenement, he is taxed in a variety of ways, and the income of the ground and land or house lord must be made up from his portion of rent; thus, if he pay 1 s. per week, or 2 *l.* 12 s. per year, he pays a proportion of the land-tax, of the tax on window-glass, on timber, on bricks and on various building articles, as also on the outlay of taxed labour in preparing the house, to say nothing of the window tax, or of the local rate or parish assessment, which every house must pay; it is therefore a very low estimate to say that the annual tax for all these is not less than 12 s. per annum.

My next item is on Corn and Food; the indirect effect of the corn laws in raising the price of bread and food, it is difficult to make evident by figures, but to a person who studies the subject of finance, a conclusion may be readily arrived at in his own mind; there can be no doubt that the protection which the landed interests possess enhances the prime cost of the necessaries of life by at least 20 per cent, if not more. I speak upon them collectively. While the general taxation on the labour requisite to grow and till the food still further augments the burden on the working classes; we will estimate the cost of bread and meat to a labouring man or artisan at 10 d. per diem, or 15 *l.* 4 s. 2 d. per annum, the minimum of supply to a hardworking man, who is thus indirectly mulcted, at the very least, 3 *l.* per annum.

On Clothing; the poorest clad man will require in shoes, stockings, shirts, smock-frock, trousers, hat and handkerchief, at least 60 s. worth in the year, one year with another, the taxed labour entering into the preparation and sale of these, or the duties levied on the importation of the raw materials, add a sixth to the cost ere they reach the consumer, who thus pays a tax of 10 s. per annum.

. . . Total taxes on the labourer per annum, 11 *l.* 7 s. 7 d. Taking a labourer's earnings at 1 *l.* 6 d. per diem, and computing his working 300 days in the year (which very many do), his income will be 22 *l.* 10 s.; thus it will be admitted, that at the very least, half of his income is abstracted from him by taxation; indeed this is rather a low estimate, for do what he will, eating, drinking or sleeping, he is in some way or other taxed. I therefore concur in the opinion which has been ascribed to the Lord Chancellor of England, that the poor man is taxed from his cradle to his grave. . . .

The effect produced on the trade on these articles by the effect of such heavy taxes upon them is, to diminish it in an extraordinary manner, and not merely diminish the home trade, but our colonial trade and our foreign trade. . . .

The middle classes, which form, I would say, the most valuable portion in the community of this country, are grievously oppressed by the present system; many of them are sinking or living upon the principal of their little property, eating, in fact, into their capital. But there is another evil arising from this system, and that is, the labouring classes are not enabled to add to the wealth of the State, to the revenue of the State, by reason of heavy taxation, which precludes their using articles on which every individual, even the poorest, would pay some proportion, which he is not now enabled to do. . . . The incidence of all taxes fall sooner or later on the consumer; but at present it bears so directly on the consumer, without intervening through the property of the country, that the poor who consume the minimum of the necessaries of life have the greatest burthen.

In speaking of the situation of the lower classes, I assume that their condition would necessarily be improved, together with their moral condition and their better feeling, if their means of subsistence were increased.

To give an illustration of the present unequal system of taxation; we tax the hand-loom weaver very highly, but the machine, or the capital which sets that machine in operation, is almost untaxed; therefore, if you tax the hand-loom weaver who throws his shuttle, you should tax the capital which competes with him, or the properties derived from such competition. I advert in that answer to the power-loom, as also to every other species of machinery which comes in competition with human labour. . . .

On the Poor Laws

In 1817 and 1818 debates about whether the current Poor Laws should be reformed took place inside and outside Parliament. Some people argued that the Poor Laws had negative consequences for the poor and society as a whole, while others argued that they were essential in a time of economic hardship.

In the included section of the Select Committee on the Poor Laws, we find several Malthusian themes: the claim that the character of the British working class is harmed by the public provision of shelter and/or money, and evidence that the costs to support the poor are rising significantly over time, as Malthus would have predicted. What evidence do the authors of the Select Committee give for thinking that the Poor Laws are counterproductive in these ways? Would your character agree that the Poor Laws have these effects, or regard them as reasons to modify or abolish the Poor Laws? Despite these negative consequences, what would your character think about the benefits of the Poor Laws for the country as a whole?

To fully consider this last question, it is important to look at the second supplied reading, from Thomas Courtenay. What salutary effects does Courtenay see as a result of the operation of the Poor Laws?

Testimony to Parliament about the Poor Laws

Source: Great Britain, Parliament, House of Commons, Select Committee on the Poor Laws. *Report from the Select Committee on the Poor Laws with the Minutes of Evidence Taken Before the Committee. Ordered by the House of Commons* (London: C. Clement, 1817). *Note:* The original has been altered to include paragraph breaks.

. . . This statute enacts, that "the churchwardens and overseers" shall take order from time to time (with the consent of two or more justices) for setting to work the children of all such whose parents shall not be thought able to keep and maintain their children; and also for setting to work all such persons, married or unmarried, having no means to maintain them, and use no ordinary or daily trade of life to get their living by; and also to raise by taxation, &c. "a convenient stock of flax, &c. to set the poor on work"; and also competent sums of money for and towards the necessary relief of the lame, impotent, old, blind, and such other among them, being poor and not able to work. . . .

But such a compulsory contribution for the indigent, from the funds originally accumulated from the labour and industry of others, could not fail in process of time, with the increase of population which it was calculated to foster, to produce the unfortunate effect of abating those exertions on the part of the labouring classes, on which, according to the nature of things, the happiness and welfare of mankind has been made to rest.

By diminishing this natural impulse by which men are instigated to industry and good conduct, by superseding the necessity of providing in the season of health and vigour for the wants of sickness and old age, and by making poverty and misery the conditions on which relief is to be obtained, Your Committee cannot but fear, from a reference to the increased numbers of the poor, and increased and increasing amount of the sums raised for their relief, that this system is perpetually encouraging and increasing the amount of misery it was designed to alleviate, creating at the same time an unlimited demand on funds which it cannot augment; and as every system of relief founded on compulsory enactments must be divested of the character of benevolence, so it is without its beneficial effects; as it proceeds from no impulse of charity, it creates no feelings of gratitude, and not unfrequently engenders dispositions and habits calculated to separate rather than unite the interests of the higher and lower orders of the community; even the obligations of natural affection are no longer left to their own impulse, but the mutual support of the nearest relations has been actually enjoined by a positive law, which the authority of magistrates is continually required to enforce. . . .

The result however appears to have been highly prejudicial to the moral habits, and consequent happiness, of a great body of the people, who have been reduced to the degradation of a dependence upon parochial support; while the rest of the community, including the most industrious class, has been oppressed by a weight of contribution taken from those very means which would otherwise have been applied more beneficially to the supply of employment.

And, as the funds which each person can expend in labour are limited, in proportion as the poor rate diminishes those funds, in the same proportion will the wages of labour be reduced, the immediate and direct prejudice of the labouring classes; the system thus producing the very necessity which it is created to relieve. For whether the expenditure of individuals be applied directly to labour, or to the purchase of conveniences or superfluities, it is in each case employed immediately or ultimately in the maintenance of labour. . . .

What might have been the amount of the assessments for the poor during the 17th or 18th centuries, the Committee have no means of ascertaining. . . . it was not till the present reign in the year 1776, that authentic accounts of this expenditure were required under the authority of the legislature. From the Returns made under Acts passed in that and subsequent years, it appears that in 1776, the whole sum raised was 1,720,316 *l.* of which there was expended on the poor, 1,556,804 *l.*; on the average of the years 1783, 1784 and 1785, the sum raised was 2,167,749 *l.* expended on the poor 2,004,238 *l.*; in 1803 the sum raised was 5,348,205 *l.* expended on the poor 4,267,965 *l.*; in 1815, 7,068,999 *l.* expended on the poor 5,072,028 *l.* . . . it is apparent that both the number of paupers, and the amount of money levied by assessment, are progressively increasing; while the situation of the poor, appears not to have been in a corresponding degree improved; and the Committee is of opinion, that whilst the existing poor laws, and the system under which they are administered remain unchanged, there does not exist, any power of arresting the progress of this increase, till it shall no longer be found possible to augment the sums raised by assessment. . . .

What number of years, under the existing laws and management, would probably elapse, and to what amount the assessments might possibly be augmented, before the utmost limitation would be reached, cannot be accurately ascertained; but with regard to the first, Your Committee think it their duty to point out, that many circumstances which, in the early periods of the system, rendered its progress slow, are now unfortunately changed.

The independent spirit of mind which induced individuals in the labouring classes to exert themselves to the utmost, before they submitted to become paupers, is much impaired; this order of persons therefore are every day becoming less and less unwilling to add themselves to the list of paupers.

The workhouse system, though enacted with other views, yet for a long time acted very powerfully in deterring persons from throwing themselves on their parishes for relief; there were many who would struggle through their difficulties, rather than undergo the discipline of a workhouse; this effect however is no longer produced in the same degree, as by two modern statutes the justices have power under certain conditions to order relief to be given out of the workhouses, and the number of persons to whom relief is actually given, being now far more than any workhouses would contain, the system itself is from necessity, as well as by law, materially relaxed.

In addition to these important considerations, it is also apparent, that in whatever degree the addition to the number of paupers depends upon their increase by birth, that addition will probably be greater than in past times, in the proportion in which the present number of paupers exceeds that which formerly existed; and it is almost needless to point out, that when the public undertakes to maintain all who may be born, without charge to the parents, that the number born will probably be greater than in the natural state. . . .

There are many modes by which the compulsory application under the provisions of a statute, of the funds which provide the maintenance of labour, would tend most materially to place the labouring classes, in a much worse condition than that in which they would otherwise be situated.

1st. An increased demand for labour is the only means by which the wages of labour can ever be raised; and there is nothing which can increase the demand, but the increase of the wealth by which labour is supported; if therefore the compulsory application of any part of this wealth, tends (as it always must tend) to employ the portion it distributes less profitably than it would have been, if left to the interested superintendence of its owners, it cannot fail by thus diminishing the funds which would otherwise have been applicable to the maintenance of labour, to place the whole body of labourers in a worse situation than that in which they would otherwise have been placed.

2dly. The effects of holding out to the labouring community, that all who require it shall be provided with work at adequate wages, is such as to lead them to form false views of the circumstances in which they are likely to be placed; as the demand for labour depends absolutely on the amount of the wealth which constitutes its support, so the rate of wages can only be adjusted by the proportion that demand bears to the supply. Now it is on the greater or less degree of nicety in which that supply is adjusted to the demand, that the happiness of the labouring classes absolutely depends.

If the demand for labour increases faster than the supply, high wages are the natural result; labourers are enabled to provide better for their children; a larger proportion of those born are reared; the burthen, too, of a large family is rendered lighter; and in this manner the marriage and multiplication of labourers are encouraged, and an increasing supply is enabled to follow an increased demand. If, on the contrary, the waste or diminution of wealth should reduce the demand for labour, wages must inevitably fall, and the comforts of the labourer will be diminished, the marriage and multiplication discouraged until the supply is gradually adapted to the reduced demand. It is obvious, that the condition of a country, whilst this latter course is in progress, must be painful; but it is more or less so according to the degree in which the foresight of the industrious classes might have prepared them for such a reverse. The habits and customs of the labouring classes in different countries must in a great degree depend on the circumstances which, by affecting the demand for labour, regulate the condition in which they are content to exist.

But where prudent habits are established, they avail themselves of a high rate of wages, to better their condition, rather than greatly increase their numbers. In England a labourer would not, formerly, have thought himself justified in marrying unless he had the means of providing himself with many things which in other countries would have been deemed unnecessary luxuries. In a state similar to this, if the labouring classes are met by a fall in wages, they will always have something to spare, which will assist in mitigating any difficulties to which they may be exposed.

Though it is by contemplating the possibility of a reverse that they can alone be stimulated to prepare for it, it is, unfortunately, far less difficult to induce men to neglect all such preparation; by holding out to the labouring classes, that they shall at all times be provided with adequate employment, they are led to believe they have nothing to dread while they are willing to labour.

The supply of labour, therefore, which they alone have the power to regulate, is left constantly to increase, without any reference to the demand, or to the funds on which it depends. Under these circumstances, if the demand for labour suddenly decreases, the provisions of the poor law alone are looked to, to supply the place of all those circumstances which result only from vigilance and caution; the powers of law, whilst they profess to compel both labour and wages to be provided, under these circumstances, in reality effect nothing but a more wasteful application of the diminished capital than would otherwise take place; they tend thereby materially to reduce the real wages of free labour, and thus essentially to injure the labouring classes.

In this situation of things, not only the labourers, who have hitherto maintained themselves, are reduced, by the perversion of the funds of their employers, to seek assistance from the rate, but the smaller capitalists themselves are gradually reduced, by the burthen of the assessments, to take refuge in the same resource.

The effect of these compulsory distributions is to pull down what is above, not to raise what is low; and they depress high and low together, beneath the level of what was originally lowest. If these views of the effect of undertaking to provide employment for all who want it are founded in truth, there results from them an obvious necessity of abandoning gradually the impossible condition, that all who require it shall be provided with work, which, whether or not, it be the real object of the statute, has by many been held to be so.

On this head, Your Committee submit, that if the provision which they have pointed out be made for children whose parents cannot maintain them, and the provision also for such as are of the class of poor and impotent be continued, the labouring classes will continue to be relieved from the heaviest part of their necessities. . . .

As whatever money would have been applied to the maintenance of these persons by the means of the poor rate, cannot fail to be employed in some such way as to put other labour in motion, the money thus restored to its natural channel cannot fail to assist in increasing the natural demand for labour; and if the wages of agricultural labour, which are now in so great a proportion paid through the poor rate, were, left to adjust themselves by the operation of the market, it could hardly fail to have the effect of gradually raising the wages of labour; for it is the obvious interest of the farmer that his work should be done with effect and celerity, which can hardly take place unless the labourer is provided according to his habits, with such necessaries of life as may keep his body in full vigour, and his mind gay and cheerful. . . .

Summary of a Report to Parliament about the Poor Laws

Source: Great Britain, Parliament, House of Commons, Committee on the State of the Poor and of the Poor Laws, and S. W. Nicholl, *A Summary View of the Report and Evidence, Relative to the Poor Laws, Published by Order of the House of Commons, with Observations and Suggestions* (York: Printed for W. Alexander, 1818).

On a compulsory Provision for the Poor.

The distresses of the last two years, have brought the situation of the Poor and the tendency of the Poor Laws, very prominently before the country; and a general opinion has arisen or been confirmed, that those laws are equally inefficient and mischievous; that they create the poverty they were meant to relieve, and impose an intolerable and augmenting burden on one portion of society, without materially adding to the comforts of the other. The late Committee has adopted this sentiment in its strongest form; and deems this principle of evil, so radically and progressively inherent in the system, that the time must come, when the profits of the land will be absorbed in the demands of the Poor; and when all efforts of cultivation must be abandoned. . . .

I am not at all disposed to consider legal relief as a right of the Poor; it is even at this period established in very few countries; and has probably been established in none for many centuries. Paley says, with his usual terseness of manner: "They who rank pity amongst the original impulses of our nature, rightly contend, that when it prompts us to the relief of human misery, it indicates sufficiently the divine intention and our duty." The general duty of charity will no doubt be admitted; the particular mode of performing that duty, is not prescribed to us; nations, like individuals, must act on their own discretion. The great body of the Poor, have no more distinct claim on the property of the country at large, than any single pauper has on any private fortune. Municipal law may give or refuse, as individual will may offer or withhold.

The poor ought to be informed of this; they ought

to know that they are in the enjoyment of a *bounty*, not in the perception of a *right*. It might be impossible to give them very accurate views of the distinction; but they would easily understand, that the provision established by law here, was almost unknown in any other part of the world; and that what the law, and nothing but the law gave, the law could also wholly withdraw. . . .

The Parish rate offers a bonus to fraud—to extravagance and to idleness; where money is to be obtained by pretended wants, wants will be pretended; where to the wages of labour parish pay is added, those wages will not be husbanded with that care which would be shown, were the wages the only source of subsistence; and many will not be over solicitous in finding work, when they can get relief independent of their own exertions. . . .

The two charges against the Poor Laws, that they tend universally to relax the exertions of the labourer, to substitute dependance on the rate, to dependance on the wages of industry, to set Parochial allowances in systematic and premeditated opposition to the produce of personal exertions; and that they are burdening the country with an intolerable and increasing pressure of taxation, are so serious, so truly alarming, that they call for unlimited abolition of the laws, if true; and full refutation if founded in error. . . .

The real evil is probably this, that in the Poor there is no system at all—no premeditation—no forethought. They act for the day without thinking of the morrow. . . .

Be the demerits of the Poor Laws what they may, they possess counterbalancing advantages of no inferior value. Amongst these the increase of a healthful and happy population stands first. Allowing that fraud is in many cases successful, and indolence in others encouraged, on the whole a large proportion of the labouring classes of this country are sober, decent, orderly, possessed of many comforts, and content with their station.

That we do not see our own blessings, is a truth not solely applicable to the state of the Poor; but others see them for us; and there is no one circumstance more marked by the foreigner, than the means of enjoyment possessed by the lower orders of this kingdom. Suppose we could not absolutely connect this state of things with the Poor Laws as an effect proceeding from them; yet, as it has clearly arisen whilst these laws were in full operation, it would be bold indeed to determine the contrary; that is, absolutely to deny all connection. I do not hesitate very much to attribute the comforts and respectability of the labouring class, to the general and compulsory right of maintenance, in precluding or at least greatly checking a system of vagrancy. It has given to the poor, that character which we only see; but which those who are acquainted with the poor of other countries, view with admiration.

When from the reverses to which commerce is at all times liable, ten, twenty, thirty thousand inhabitants of a single town, are reduced to distress, what, short of compulsory support, could ward off the horrors of famine? Without this preventative, the most dreadful diseases would arise in one generation, and be communicated to others. Vagrancy must become almost universal; for it is only by securing subsistence to the pauper at home, that you can pretend to exclude him from seeking it abroad; and than vagrancy, there is no more determined enemy of health, morals, and industry.—The plan might commence with the inhabitants of towns in a state of decay; but the idle, the profligate, and the dishonest of other places, would join them; and so the practice would spread without limit.

Even amongst the well disposed poor, cases of distress from illness or the pressure of a family, must at all times be numerous; and where now, occasional relief retains a man in his village, the want of that relief would throw his whole family into a state of mendicity. At first his children would apply in their immediate neighbourhood, next the circle would be somewhat enlarged, the mother would then join the party, and at length the whole family would take their station with the permanent and hardened beggar.

The Poor Laws have unquestionably checked the system of Vagrancy in a very great degree. If they do not hereafter extirpate it, the fault lies with the Constable, and the Magistrate. When every man has a sure subsistence at his own home, it is a culpable indolence, and not a feeling of humanity, which permits him to wander as a vagrant elsewhere. Though the vagrant when stationary adds somewhat to the burden of his particular parish, do not the united theft and extortion to which his vagrancy gives rise, add ten times more to the burden of the kingdom at large! . . .

Thomas Courtenay on the Poor Laws

In this work Thomas Courtenay is responding to the House of Commons Committee on the State of the Poor and of the Poor Laws. In the first few pages he disagrees with the committee's statement that the Poor Rates will be too expensive for those with property to afford. He admits that more people are relying on the Poor Law than previously, but he thinks the committee exaggerates the extent to which the Poor Law encourages an unsustainable increase in population. He also does not believe that the increase in Poor Rates will lead to the destruction of the property-owning classes.

Source: Thomas Peregrine Courtenay, *A Treatise upon the Poor Laws* (London: J. Murray, 1818).

. . . It has been said to be unjust to tax property for the relief of indigence; that such a system is fundamentally inconsistent with the order of Society of which the Law of Property is a sacred and essential branch, and that it leads to an Equality of Condition, and a Community of Goods. No man can be more satisfied than I am of the utter impracticability of all Schemes of Equality.

If a systematic relief of the indigent by the more wealthy, tend to this result, it is confessedly vicious. But of this tendency there must be reasonable proofs. . . .

It is surely, in the abstract, reasonable and just, that the excessive inequalities of condition which Society occasions among men, should be corrected by a sacrifice on the part of the more prosperous, sufficient to preserve from wretchedness or actual destruction the least fortunate classes. That would doubtless be the most perfect state of Society, in which no such abject inferiority should exist. . . . it is assuredly neither absurd nor unnatural to attempt, as well in our social institutions, as in our individual conduct, the mitigation of the severe but necessary evil. . . .

I readily admit, that the system, as now administered, does press too severely upon the class immediately above those who receive relief; and that it has, so far, the tendency imputed to it in the Report "to pull down what is above"; but no argument drawn from the inequality of the pressure, is conclusive against systematic relief. . . .

I confess, that regarding the Poor Laws as part of a political system which has on the whole been prosperous and successful, and as founded upon a principle of apparent benevolence, I think it natural to hesitate in abandoning them. But in truth, much that has been now said favourable to the Poor Laws, has been excited by what appeared to me exaggeration in condemning them. I am far from denying, that that state of Society would be the best, in which there should be the utmost degree of attainable comfort, in all classes, without any compulsory assistance of one by the other; nor am I insensible to the many certain, severe, and crying evils which under the English system affect both rich and poor. . . .

No man ought to make up his mind to the abolition of the whole code of Poor Laws, without satisfying himself of the truth of one or other of the following propositions:—

1. That miserable poverty will not occur.
2. That occurring, it will be relieved by private benevolence.
3. That it ought not to be relieved, but left to operate as a punishment or as a warning.

And to reconcile himself to the simultaneous abolition of every part of the code, he must hold some one of these propositions as applying with equal certainty to each of the various cases and qualities of misery, which are relieved under the Poor Laws. He must convince himself that under his reformed system every single case of helpless want will either be prudently avoided, privately relieved, necessarily abandoned, or worthily punished. . . .

The Poor Laws are, in their operation, attended with much evil; they press severely upon property, and they engender bad habits among the people. Yet, they have not prevented England from growing in wealth and strength, and when we accuse them of breaking the spirit of the people, we forget that it is while this bad system has been in full activity amongst our peasants and artisans, that our armies, recruited among them, have been growing in renown for bravery and perseverance.

There may be, and doubtless are, grievous faults in our economical as in our political institutions; but it is hard to say that either is incurably vicious, seeing how, with these faulty institutions, we have stood through the convulsions of the last five and twenty years. We may have stood, notwithstanding, rather than by the force of, our system of polity; but if we recollect, at the same time, the peculiarity of our institutions, and the singularity of our fate, it is not unnatural to imagine that the one may be connected with the other. If "our timber has thriven in its strength of trunk, and pride of branch and foliage," we may justly conclude that the soil is good, and the roots unhurt.

. . . I will plainly express my conviction, that although there is much in our system which requires amendment, we have no reason for apprehending that the Poor Laws will destroy us; and that, still retaining in our code, the principle of national charity, we shall not cease to be eminent among the nations of Europe, for the freedom and stability of our Constitution, the variety of our Resources, or the strength of our People.

On Political Reform

The following are newspaper reports describing reform meetings and the speeches of some key reformers, especially William Hunt. They outline some of the key ideas of those who wanted reform and how they went about trying to achieve it.

William Hunt argues that one of the causes of destitution among the working classes is excessive taxation, that corruption is endemic among the ruling classes, and that wealthy classes look out for their interests to the exclusion of the interests of the poor. Would your character agree or disagree with Hunt's claims? Why?

What would your character think of Hunt's solutions? Would she regard them as but a prelude to the Jacobin excesses of the French Revolution, as some conservatives did? What would your character propose regarding the political structure of Great Britain?

Opposition to universal suffrage (extending the right to vote to all males, or, more controversially, to all British subjects) was common in the early nineteenth century. John Horne Tooke gives some of the popular arguments against universal suffrage in our final reading in this section. One of the key premises in his argument is the claim that an agent's power in a system ought to be proportional to his contributions. Tooke's argument presupposes that power is a function of financial ability to contribute to the administration of the state. Would your character agree that power in a system ought to be proportional to one's ability to contribute? What factors would your character consider most important when evaluating an individual's ability to contribute to the health of the state?

Times (London) Report of a Reform Meeting

Source: "Meeting at Spa-Fields," *Times* (London), 16 November 1816, 2.

In consequence of an advertisement which was placarded throughout the metropolis, stating that a meeting of manufacturers, artisans, &c. would be convened in these fields, to take into consideration the propriety of petitioning the Prince Regent upon the present distressed state of the country, an immense concourse of people was yesterday assembled, to the number of about 5 or 6,000 persons. The meeting was advertised to be held at 12 o'clock, for 1; but long previous to that hour crowds were seen to flock from all parts to give their attendance.

At about half past 12, a hackney-coach, containing four persons, was seen to drive into the fields, pelted with dirt and mud by the mob. Upon his arrival in the midst of the crowd, Mr. PARKES addressed them from the window of the coach, and requested them to be tranquil. . . .

[Parkes was not one of the intended speakers but he gave a speech that the crowd appreciated.]

The meeting adjourned, after the speech of Mr. Parkes, to a public-house called Merlin's Cave, in the neighbourhood, where two persons who signed the requisition were attending the arrival of Mr. Hunt. A long pause ensued, during which many symptoms of impatience were exhibited both in and out of the house. About one o'clock the orator appeared in the centre of the multitude, and was conducted amidst their acclamations to a hackney-coach, on the roof of which, after the example of his predecessor, he mounted, and addressed a few sentences to the populace. Feeling, however, some inducement to occupy a more advantageous station, he soon abandoned this post, and gratified the committee men with his presence in the ale-house. His approach was preceded by a three-coloured flag, and a cap hoisted to a pole. Here he recommenced his harangue from one of the windows of the front room, in which large libations had been already offered on the altars of patriotism. . . .

They had met there for the purpose of petitioning the Prince Regent and the Legislature for some effectual relief to those growing miseries, the tale of which would require a month to tell, and a month fully to understand. He heard only yesterday of a poor labourer of Spitalfields, who, with a wife and three children, had been heard to pray that some friendly hand would deliver him from the intolerable load of life. What, then, was the cause of this unparalleled and universal distress? What is, as their enemies had the impudence to promulgate, the indolence and improvidence of the poor? This was the cause assigned by a corrupt and hireling press; but it was untrue. On the face of the earth there was no people so industrious as the people of this country. An Englishman worked as much in a week as any other labourer in a month. *(Applause, and cries of, "We do, we do.")* What was the cause of the want of employment?—Taxation. What was the cause of taxation?—Corruption. It was corruption that had enabled the borough-mongers to wage that bloody war which had for its object the destruction of the liberties of all countries, but principally of our own. Its object had been confessed by Ministers themselves to be the restoration of despotism abroad, and persecution at home. They knew that persecution alone would enable them to reduce the people to that state of feebleness in which they could no longer resist. With both these views, therefore, they began by re-establishing the Bourbons in France, and in Spain, and the Pope with his infernal inquisition in Italy. . . .

They had now in their view the British Bastile [*sic*] (pointing to the Cold-bath-fields prison), where so much tyranny had formerly been exercised, and to which so many miserable victims had been consigned. . . . He begged them to beware of false friends, of wolves in sheep's clothing.

[Hunt complained of those "patriots" who did not attend this meeting to speak to the crowd because they were afraid.]

He scrupled not to say that they were the greatest enemies the people ever had. Two years ago, when it was proposed to petition for the repeal of the property tax, he had proposed that, as they were about to petition for the relief of the rich, something should be added with a view to the relief of the poor. *(Applause.)* The rich man, he undertook to prove, who brewed his own beer, had it 10s. a barrel cheaper than the useful and industrious mechanic, who was obliged to buy it at the public-house. Up started one of the wolves in sheep's clothing, and abused him as hawking about his speeches in all parts of the country. And what disgrace, he desired to know, was there in endeavouring every where to remedy the sufferings of his fellow-countrymen? *(Applause.)* This wolf, this hireling of the Government, opposed his motion; and why? Because the miseries of the poor were altogether out of their consideration. The rich alone were to be relieved, and the poor might go to the devil their own way. . . .

[*Hunt criticized those who compromise with the government.*]

He himself had never had any connexion with Government, nor, as it was a present constituted, would he ever have. He never should be found in any dirty scrapes, by which he might be tempted to sacrifice his political principles. They had two sorts of enemies to encounter—one class who preferred open hostility, because they were the immediate offspring of corruption; another, more insidious, who called themselves friends, and came there for the purpose of creating some ridiculous disturbance, which might serve as a pretence for calling out the military. It was their business to frustrate the designs of both parties, and to show them of what stuff Englishmen were made. All ranks, save the children of corruption who fattened on the vitals of the country, were alike involved in one common distress. It was not alone the artisan, with a dirty apron round his waist; it was the man who wore a good coat, who had to bear his equal portion of the national suffering. They ought to be the support of each other; but how could he find employment for the poor, if all his means were exhausted by taxation? He well knew what ought to be done in such a crisis. He knew the superiority of mental over physical force; nor would he counsel any resort to the latter till the former had been found ineffectual. Before physical force was applied to, it was their duty to petition, to remonstrate, to call aloud for timely reformation. Those who resisted the just demands of the people were the real friends of confusion and bloodshed; but if the fatal day should be destined to arrive, he assured them that, if he knew any thing of himself, he would not be found concealed behind the counter, or sheltering himself in the rear. Every thing that concerned their subsistence or comforts was taxed. Was not their loaf taxed—was not their beer taxed—were not their coats taxed—were not their shirts taxed—was not every thing that they ate, drank, wore, and even said, taxed? *(A laugh.)* What impudence, what insolence was it then in the corrupt and profligate minions of Government to say that the people suffered nothing by taxation. If there were not taxes, the labourer would have his quartern loaf for 4d., his pot of beer for 2d., his bushel of salt for half-a-crown, his soap, candles, sugar, tea, and other articles, for half their present price. It was necessary to have some taxes, he allowed; but every dictate of justice, every right of the people, called for their reduction. They were imposed for no purpose in which the nation was interested. They were imposed by the authority of a boroughmongering faction, who thought of nothing but oppressing the people, and subsisting on the plunder wrung from their miseries. *(Applause.)*. . . .

[*Hunt criticized the use of tax money to pay for pensions and jobs for relatives of members of Parliament and the British royal family.*]

All parties were pensioned, whether Whigs or Tories, whether the Pitt or the Fox faction, whether ins or outs; all fattened themselves on the spoils of the people, all were tarred with the same brush. *(Loud applause.)*. . . .

This put him in mind of a story (which he had often repeated before) of a robber who stripped a traveller of a thousand pounds, and took a merit to himself for returning him a penny to pay the turnpike; or of a thief who stole a goose, and returned the giblets. When they gorged themselves with the public spoils, and subscribed a paltry sum for the relief of those who, without their oppressions, would be happy and affluent, they called their conduct benevolent, and vilified those who did not give them credit for disinterestedness and public spirit. The press had been employed to vilify him and the friends of liberty. . . .

The meeting whom he now addressed had assembled manfully to express their opinions and complaints in a legal and constitutional manner, and not for rioting or disturbance. Their enemies wished for some pretence for calling in the military; and some of the milk sop patriots had been deterred from attending, lest that should happen. He was not moved by any such terror—at the risk of his life he would perform his duty. . . .

It was the duty of every Englishman to petition for a reform in Parliament. . . .

He concluded with moving the resolutions, which were to the following effect:—

That the country was in a state of the most fearful and unparalleled distress and misery, felt by all classes excepting those who derived their fortunes from the taxes levied upon the people; the farmer, the manufacturer, and the tradesman, were all involved in the same lamentable oppression.

That the cause of these intolerable burdens was, first, the immense amount of debt contracted by the borough-mongers, for the purpose of carrying on a long, and unnecessary, and unjust war; the main object of which appeared to have been to stifle civil, political, and religious liberty, and to restore despotism throughout the country; secondly, the maintenance of an army in France against the unanimous wish of the French nation; thirdly, the keeping up of an enormous standing army, with a view to curb the people, and to compel them to submit to pay war-taxes in times of peace: and, fourthly, the lavish expenditure of the public money by innumerable men and women who hold pensions, grants, and annuities, without performing the smallest service to their country.

That the sole cause of these abominable practices was a want of reforms in Parliament, and by the return of members to the House of Commons by such means as were confessed to be by one of its members as notorious as the sun at noon-day.

That a petition be presented to his Royal Highness the Prince Regent, beseeching him to take into his consideration the burdens of this suffering, and patient, but starving people; and imploring his Royal Highness to cause Parliament to be assembled, in order that measures might be adopted to redress these evils, to feed the hungry and to clothe the naked, so that the unhappy and starving people might be preserved from desperation; and, above all, to listen, before it was too late, to the earnest and repeated prayers of the nation. . . .

Times *(London)* Editorial against "Radical" Reform

Source: A Friend to X, "To the Editor of the Times," *Times* (London), 18 November 1816, 2.

SIR—Amongst a people so enlightened as the English of the present time, there is nothing very formidable in the direct operation of Jacobinism. The Jacobin spirit, once clearly detected, is sure to be overthrown by the resolution and intelligence of that great bulk of the nation who consider the protection of the rich man's property essential to the subsistence of the poor, and who mean by liberty the enjoyment of a universal security against outrage or violence from any quarter—from King, Magistrate, or mob. As for patriots and legislators like orator HUNT, they are not fit subjects for admonition: the end of their course will be the best explanation of their principles, and their final address from the cart of the executioner the most apt commentary upon those virtuous doctrines with which, from the top of their vehicles, they now edify and animate the same admiring audience. It has always been a fashion with that wholesale mobmonger to take advantage of the empty stomach of his hearers, and after haranguing them for several hours, until they are half dead with hunger and thirst, he descants on the blessings of a quarten loaf for 4d, and a pot of beer for 2d; when some hundreds of poor desperate wretches run away to gut the butchers' and bakers' shops, and more than realize the promises of their master by laying in provisions at no higher price than the risk of transportation or the gallows. . . .

[The author does not fear for the persons or interests against which Hunt directs his criticisms.]

No, Sir my fears are on behalf of that cause of which the factious incendiary professes himself an advocate. I cannot forget that it was the vexatious turbulence and disorder of the Republican government of this country which prepared men's minds to submit to the usurpation of CROMWELL, nor that the desolating anarchy and irregular fury of the French Jacobins laid the foundation for the military government, which, however harshly exercised by BUONAPARTE, his subjects, worn out by successive persecutions, regarded less as the grave of freedom than the refuge of individual safety and repose. The English Jacobins of 1793 we owe some, though less signal obligations, than did our ancestors to the Roundheads of the 17th century or the well disposed among the French people to the Terrorists FOUCE, ROBESPIERRE, AND MARAT. . . .

[*The author blames those English people who supported the French Revolution in 1793 for the suspension of liberties in England that the government felt was necessary at the time—i.e., "Pitt's reign of terror"*]

Undoubtedly, Sir, I am not one of those who question the necessity of the war of 1793. While the British Constitution should remain, it stood a rebuke to the democratic visions of the Revolutionists; while this remained an independent empire, the military dreams of France were equally absurd; war was therefore declared against Great Britain in both these hated characters, as an example of temperate liberty, and as a check to inordinate ambition. We could not help the war, I repeat, since, for a period of 15 years, it involved our existence as an unconquered people. But it is permitted us to regret a multitude of ill consequences that arose out of the feeling with which the war was commenced. We felt that we had a constitution to protect; but we were often terrified, or cheated into a belief, that the constitution was no longer capable of affording us protection. . . .

[*The people of England began to fear that the English Jacobins had influenced the military and were going to take over the country, so they gave Pitt's Ministry, and the monarch, more power than ever before. This new military and monarchical spirit threatened the old constitution. The monarch's new power also led to an extension of patronage, the multiplication of offices and salaries, and thus to an increase in the expenses of the government.*]

[*These were some of the*] . . . injuries we have sustained from the revulsion of public sentiment produced by the savage extravagance of English Jacobins. . . .

[*There is an army of 25,000 men kept in England for the first time during a period of peace, but it cannot be dismissed*] " . . . when Mr. HUNT and his rebellious crew talk openly of attacking and overthrowing the State?" Thus does the spirit of unconstitutional democracy obstruct and harass the real friends of liberty and of the constitution: this, which seems to be a breathing time sent from Heaven to revise our conduct—to correct our errors—to redress our grievances—to restrain our rulers—and to regain some portion of that precious ground which our laws, our finances, and our rights have lost—this is the moment seized upon by the evil genius of Jacobinism to disgust the proud patriot—to affright the timid—not only to confirm and almost to make hereditary the possession of the government by the same class of politicians who have already held it *for two-and-thirty years*, but to give them a plea for continuing the most mischievous parts of their domestic policy, and for still farther extending the most dangerous branches of their power.

[*The ministerial paper (i.e., Tory Party paper) has accused the Opposition (i.e., Whig Party) of aiding those people who created this confusion. The author hopes that this does not deter Whigs*] . . . from prosecuting those measures of practical reform, without carrying which we shall have really no alternative between an arbitrary Monarch and a more arbitrary mob. I trust, therefore, that the magistracy will be active and resolute. I trust the gentlemen of England will gallantly stand forth to quench the lawless firebrand themselves, and not meanly seek protection from the bayonet and sabre, which, once acknowledged as the supreme preservers of the public peace will grant us no peace but at the expense of our dearest liberties. If indeed, a despotic Minister were to choose his weapon for the ruin and subjugation of a free people, he need devise no better than a knot of Jacobins to make the very name of freedom odious, and to intimidate or discredit every patriot in the land. It is for the friend of the constitution of England to meditate deeply the subject of this letter; for head and heart will both be exercised by the double duties which await them.

Editorial to the Times (London) against Universal Suffrage

Source: W. J., "To the Editor of the Times," *Times* (London), 15 February 1817, 3.

SIR,—The turbulence of some of the universal suffrage gentlemen may be somewhat allayed by reading the following extract from the letter of a gentleman who, all parties will agree, was full as competent to form an opinion on the subject of reform as any of our modern speculators. The letter alluded to was written by the late Mr. Horne Tooke to Lord Ashburton, the 10th of May, 1782. . . . *[Tooke was replying to arguments for universal representation made by Major Cartwright]* . . . who asserts, that "No man can be free who has not a voice in framing those laws by which he is to be governed. He who is not represented has not this voice; therefore, every man has an equal right to representation, or to a share in the Government. His final conclusion is, that every man has a right to an equal share in representation." Upon which opinion Mr. Tooke makes the following remarks:—"Now, my Lord, I conceive the error to lie chiefly in the conclusion; for there is very great difference between having an equal right to a share, and a right to an equal share. An estate may be devised by will amongst many persons, in different proportions; to one 5 *l.*, to another 500 *l.*, &c.; each person will have an equal right to his share, but not a right to an equal share."

"This principle is farther attempted to be enforced by an assertion, that "the all of one man is as dear to him as the all of another man is to that other." But, my Lord, this maxim will not hold by any means; for a small all is not, for very good reasons, so dear as a great all. A small all may be lost, and easily regained; it may very often, and with great wisdom, be risked for the chance of a greater; it may be so small, as to be little or not at all worth defending or caring for. *Ibit eo qui zonam perdidit.* But a large all can never be recovered; it has been amassing and accumulating for many generations; or it has been the product of a long life of industry and talents; or the consequence of some circumstance which will never return. But I am sure I need not dwell upon this, without placing the extremes of fortune in array against each other; every man whose all has varied at different periods of his life can speak for himself, and say, whether the dearness in which he held these different all was equal. The lowest order of men consume their all daily, as fast as they acquire it.

"My Lord, justice and policy require that benefit and burden, that the share of power and the share of contribution to that power, should be as nearly proportioned as possible. If aristocracy will have all the power, they are tyrants and unjust to the people, because aristocracy alone does not bear the whole burden. If the smallest individual of the people contends to be equal in power to the greatest individual, he too, in turn, is unjust in his demands; for his burden and contribution are not equal.

"Hitherto, my Lord, I have only argued against the equality; I shall now venture to speak against the universality of representation, or of a share in the Government; for the terms amount to the same.

"Freedom and security ought surely to be equal and universal; but, my Lord, I am not at all backward to contend, that some of the members of a society may be free and secure, without having a share in the Government. The happiness and freedom, and security of the whole, may even be advanced by the exclusion of some, not from freedom and security, but from a share in the Government.

"My Lord, extreme misery, extreme dependence, extreme ignorance, extreme selfishness (I mean that mistaken selfishness which excludes all public sense), all these are just and proper causes of exclusion from a share in the Government, as well as extreme criminality, which is admitted to exclude; for thither they all tend, and there they frequently finish." I am, your, &c

Reports of Working-Class Protest

These 1812 reports of Luddite destruction of machinery and large-scale riots in protest of the new industry give a sense of the violence that the technological changes provoked. What would your character say about these actions?

Report of Frame-Breaking in Nottingham

Source: *Times* (London), 30 January 1812, 3.

According to letters received yesterday from Nottingham, it appears that frame-breaking and other outrageous proceedings continued as usual, in Nottingham and the adjacent counties, in defiance of the large military force which has been stationed at various points for the protection of persons and property. . . . On Saturday night, two frames were broken at Bulwell. Soldiers arrived on the spot before the depredators had escaped, and several shots were exchanged; but either by good fortune or good management the rioters got clear off. About two o'clock on Sunday morning, an express arrived from Reddington, stating that 13 frames had been destroyed at that place, and that the *Luddites* were engaged in their mischievous work when the messenger came away. On Saturday and Sunday nights, not fewer than 60 frames were destroyed. *Ludd* has declared his intention of destroying all frames without exception. Five armed *Luddites* stopped a carrier about a mile from Mansfield turnpike; and having collected from his waggon every article obnoxious to them, set fire to the same. In short, the whole town of Nottingham was in the greatest alarm; the people were afraid to go to bed, as a general ransacking of the town has been promised.

We cannot close this account without again expressing our surprise at the laxity of the Government, in suffering these depredations to continue. Have they forgotten the protection which they owe to the persons and property of their Sovereign's peaceable subjects? If their remissness proceeds from any acquiescence in certain local feelings of excessive tenderness for the offenders, yet is the example most injurious to the manufacturing class of people throughout the kingdom: and more than ample time has been afforded for a gentle composure of these disturbances, which seem, though indulgence, to threaten the whole community.

Report of a Riot in Macclesfield

Source: *Times* (London), 21 April 1812, 3.

Macclesfield, April 15

The streets of this borough were on Monday last the scenes of riot.

Early in the forenoon, Mr. Daniel Rowson, a factor in the town received a letter from Stockport, threatening the immediate destruction of his and his neighbours' property. Already the rioters had begun to assemble in the fields adjoining the town; and by noon, finding they had sufficient force to carry their determination into effect, they entered the market-place. They proceeded to enquire how potatoes sold; and not approving the price, began to thrown them about the streets: five bags were thus disposed of, when the foremost of them, having a blue ribbon in his hat, was taken into custody, and lodges in the common gaol by order of the Magistrates. The word "rescue" was given, the doors of the gaol broken, the man set at liberty, and carried in triumph to the market-place in an instant of time, the mob hallooing and shouting.

After a short deliberation, two, apparently the ringleaders, held up a stick—the signal was answered by an universal huzza, and the whole body set forward to the premises of Mr. Rowson, in Mill-street. Here they demolished all the windows, broke the door of the shop, and rolled cheeses and other articles through the streets. When the destruction was complete, and another shout of triumph had succeeded, they took the road through Pickford and Sunderland-streets to the shops of Messrs. John Holland, Simon Malkin, Rupert Malkin, Samuel Clowes, Matthias Mason, and several others, some of which they completely gutted, destroying the windows and furniture, and were guilty of greater excesses as they proceeded.

A company of the Royal Cumberland Militia were mustered by beat of drum. They followed the mob from street to street without success some time; but coming up with them in Mill-street, the riot act was there read by W. Ayton, Esq. one of the Borough Magistrates. The act was repeated in the market place, without any other effect than further to exasperate the populace. The Magistrates, who had hoped to appease the tumult by a show of arms, now saw the necessity of employing force; but, still reluctant to shed blood by the use of the musket, they summoned the Macclesfield Troop of Volunteer Cavalry. The cavalry mustered at 4 p.m. under the command of their Captain, J. S. Daintry, Esq. and marched to the Waters and from thence on a hard gallop to the factory of Messrs. Goddall and Birchinall, whither the mob proceeded. The latter did not wait the charge, but crossed the river in two directions to the fields. A part took the way to Beach, the residence of C. Wood, Esq. the main body fled to the tract of waste land, that has long been the depository of the town's rubbish, and made a stand here conceiving themselves inaccessible to the cavalry. The cavalry halted in front of the main body, and the Magistrates parleyed with them to no effect for near half an hour. Meanwhile the other division attacked Mr. Wood's residence, breaking the windows and window-frames, bursting the doors, and threatening the life of Mr. Wood. The work of destruction was not finished when the Magistrates received information of it, and a part of the horse *[guard]* were dispatched to route the assailants. The main troop, with the Cumberland Militia, speedily followed; the rioters were surprised, and again retreating precipitately across the river, effected a junction with their fellows in the fields adjoining to the Manchester road. The cavalry now broke in upon them; immediate overthrow and dispersion followed; several of the ringleaders were seized, and the populace ran with loud screams in great confusion into the town. In less than ten minutes an area of 100 acres was cleared, the cavalry leaping hedges and walls in their progress, and driving all before them, amidst a storm of stones and brick-bats. The prisoners, as they came in, were delivered over to the Cumberland Militia, and marched under escort to the Town Hall, where they remained strongly guarded.

The fields having been cleared, the mob repaired to the Market-place, and adjoining streets, threatening rescue, and growing more outrageous as the night advanced. The cavalry having dismounted for refreshment, again assembled at seven; and the mob remained deaf to the persuasion of the Magistrates, orders were given to clear the streets. This was the most perilous part of the service; the rioters clung to the walls, and took refuge in the numerous alleys, throwing stones and bricks from thence with surer aim, and to the great annoyance of the troops. At length however, they were driven from these also, and reduced to their last stand in the Old Church-yard: the key of the great gates being procured, they dispersed with precipitation.

The nightly patrole of the town, 300 in number, were now collected together, and proceeded to search the inns and lodging-houses, taking many into custody for the night, and dismissing the more peaceable to their respective homes. By the hour of 11 p.m. the town was quiet; the cavalry, however, continued on duty till two, and a patrole during the night. . . .

We are now under painful necessity of recording the accidents which occurred. Of the Macclesfield cavalry, two members were severely hurt; Mr. Higginbotham, an Alderman of the Borough, had his arm broken, and Mr. Grimsditch, a solicitor, received several contusions on the head; many others were struck violently, but none dangerously wounded. Of the rioters, one person, a woman, was unfortunately trampled on, and her arm broken; the prisoner, Stubbs, also received a deep gash in the head, when in the act of violently assaulting Mr. Grimsditch.

The instigators of the riot were about 300 in number; of these, few were townsmen of Macclesfield, the majority being colliers and carters from Bellington and Rainow, or spinners from the hill country near Stockport. The total number of persons riotously assembled was at one period certainly 5,000. The boys, at the bidding of their elders, took the more conspicuous part of breaking the windows; the leaders did the indoor work.

Report of Riots in Manchester and Birmingham

Source: Times (London), 29 April 1812, 3.

MANCHESTER, April 26.

On Friday afternoon, about 4 o'clock, a large body of rioters suddenly attacked the weaving factory belonging to Messrs. Wroe and Duncroft, at Westhoughton, about 13 miles from this town; of which, being unprotected, they soon got possession. They instantly set it on fire, and the whole of the building with its valuable machines, cambrics, &c. were entirely destroyed. The building being extensive, the conflagration was tremendous. The damage sustained is immense, the factory alone having cost 6000 *l*. The reason assigned for this horrid act is, as at Middleton, "weaving by steam." By this dreadful event, two worthy families have sustained a heavy and irreparable injury, and a very considerable number of poor are thrown out of employment. The rioters appear to level their vengeance against all species of improvement in machinery. Mistaken men!—what would this country have been without such improvements? Not one of the incendiaries are taken; nor was there a solider in that part of the country.

BIRMINGHAM, April 25

The Magistrates with great satisfaction congratulate the town and neighbourhood on the restoration of public tranquility, without material injury to person or property. Their gratification has been greatly heightened by the knowledge, that none of the poor inhabitants, who may be suffering from the pressure of the times, were at all implicated in the late tumultuous proceedings, being warranted by facts in asserting, that these originated with persons of a very different description, actuated by other motives than those of relieving the distressed.

Under this impression, the magistrates now feel it to be their duty, as it is always their earnest wish and desire, to promote, as far as in them lies, the accomplishment of some effectual measures for the relief of those truly deserving persons, who not only at this time, but always, have manifested the most laudable patience under privations and hardships.

They, therefore, request, the inhabitants to meet them at the Public Office, in Moor-street, on Friday next, 1st of May, at 11 in the forenoon, for the purpose of resolving on a plan for relieving the suffering poor, by entering into subscriptions, and recommending to the higher and middle classes to lessen the consumption of bread, corn and potatoes in their houses.

It is but too true, that great pains have been taken to irritate the minds of the labouring classes, by the secret distribution of inflammatory hand-bills; but, thanks to the loyalty and good sense of our townsmen, these attempts have completely failed.

It is a curious fact that the two principal ringleaders in the late riot at Sheffield were the most ingenious mechanics in the town, and that in the week before the riot they *each* received wages for their week's work to the amount of *four guineas and an half*. We leave our readers to reflect upon the *motives* which induced these men to disturb the peace of their fellow-townsmen,—was it *want* or *wickedness*?

Report of Luddite Violence in Manchester

Source: Times (London), 23 May 1812, 3.

Letters were received yesterday from Manchester, which communicate the unpleasant intelligence, that during the last few days several violent outrages have been committed by the *Luddites*. No fewer than three persons have been shot in different parts of the country, without any discovery or even suspicions of the murderers. Besides these atrocities, a gentleman, obnoxious to the hatred of the *Luddites*, was attacked in a lane, in the middle of the day, by several men who were strangers to him: after receiving a violent blow on the head with a large stone, he had the good fortune to escape by the swiftness of his horse. A working man, who had been mistaken for another person that had given information against the *Luddites*, was taken to a coal-pit with an intent of precipitating him to the bottom, when it was discovered that he was not the man whom the assailants were in quest of; and in consequence he was suffered to depart, without sustaining any injury. All possible means have been tried to induce the *Luddites* in prison to divulge the whole extent of their plan, and to impeach their ring-leaders, but without effect. Some have been promised protection, and a competency for life, but to no purpose.

The Riot Act and Combination Acts

The Riot and Combination Acts are extremely important for understanding the legal landscape around protest and the rights of the laboring classes during the time of our game. As the working classes consider various ways of responding to events in Manchester, everyone must understand the possible repercussions of their actions. It is important to emphasize "possible" here; whether and how these laws should be enforced was contested in England at the time. Widespread opprobrium from many classes might make the authorities less willing to enforce the provisions of these acts; some may advocate for their repeal altogether.

The Riot Act provides a severe penalty—death without benefit of clergy for anyone who fails to disperse from an assembly that is riotous and unlawful after a magistrate has read to the assembly provisions from the Riot Act. It is a ban on organized protest and assembly.

The Combination Act forbids workers from forming unions ("combinations") for the purpose of increasing their wages, lowering the number of hours they are obliged to work, or lowering the amount of work they are required to do. It also forbids workers from attempting to prevent other people from filling their roles if they do go on strike. The penalty for violating the Combination Act is less severe (but still devastating for a working-class family): imprisonment for up to three months. The Combination Act does allow for "Workingman's associations," groups that may take subscriptions for educational purposes, or to pay for things like funeral expenses.

Some questions for your character to consider:

- *Does the Combination Act support or undermine the liberty of English subjects? Why? You might consider Smith's account of the asymmetrical bargaining position of laborers and workers as you develop an answer to these questions. Should workers be free to form unions and bargain as body, or is the right contract best held between two individuals—worker and master?*
- *What social instabilities would justify the invocation of the Riot Act? To what extent are they present in Manchester? To what extent does the Riot Act impede the ability of the working classes to make their views known and felt in society? Does the Riot Act advance or impede the liberty and security of British subjects?*

The Riot Act of 1714

An act for preventing tumults and riotous assemblies, and for the more speedy and effectual punishing of the rioters.

I.

Whereas of late many rebellious riots and tumults have been in divers parts of this kingdom, to the disturbance of the publick peace, and the endangering of his Majesty's person and government, and the same are yet continued and fomented by persons disaffected to his Majesty, presuming so to do, for that the punishments provided by the laws now in being are not adequate to such heinous offences; and by such rioters his Majesty and his administration have been most maliciously and falsly traduced, with an intent to raise divisions, and to alienate the affections of the people from his Majesty therefore for the preventing and suppressing of such riots and tumults, and for the more speedy and effectual punishing the offenders therein; be it enacted by the King's most excellent majesty, by and with the advice and consent of the lords spiritual and temporal and of the commons, in this present parliament assembled, and by the authority of the same, That if any persons to the number of twelve or more, being unlawfully, riotously, and tumultuously assembled together, to the disturbance of the publick peace, at any time after the last day of July in the year of our Lord one thousand seven hundred and fifteen, and being required or commanded by any one or more justice or justices of the peace, or by the sheriff of the county, or his under-sheriff, or by the mayor, bailiff or bailiffs, or other head-officer, or justice of the peace of any city or town corporate, where such assembly shall be, by proclamation to be made in the King's name, in the form herein after directed, to disperse themselves, and peaceably to depart to their habitations, or to their lawful business, shall, to the number of twelve or more (notwithstanding such proclamation made) unlawfully, riotously, and tumultuously remain or continue together by the space of one hour after such command or request made by proclamation, that then such continuing together to the number of twelve or more, after such command or request made by proclamation, shall be adjudged felony without benefit of clergy, and the offenders therein shall be adjudged felons, and shall suffer death as in a case of felony without benefit of clergy.

II.
And be it further enacted by the authority aforesaid, That the order and form of the proclamation that shall be made by the authority of this act, shall be as hereafter followeth (that is to say) the justice of the peace, or other person authorized by this act to make the said proclamation shall, among the said rioters, or as near to them as he can safely come, with a loud voice command, or cause to be commanded silence to be, while proclamation is making, and after that, shall openly and with loud voice make or cause to be made proclamation in these words, or like in effect:

Our sovereign Lord the King chargeth and commandeth all persons, being assembled, immediately to disperse themselves, and peaceably to depart to their habitations, or to their lawful business, upon the pains contained in the act made in the first year of King George, for preventing tumults and riotous assemblies. God save the King.

And every such justice and justices of the peace, sheriff, under-sheriff, mayor, bailiff, and other head-officer aforesaid, within the limits of their respective jurisdictions, are hereby authorized, impowered and required, on notice or knowledge of any such unlawful, riotous and tumultuous assembly, to resort to the place where such unlawful, riotous, and tumultuous assemblies shall be, of persons to the number of twelve or more, and there to make or cause to be made proclamation in manner aforesaid.

III.
And be it further enacted by the authority aforesaid, That if such persons so unlawfully, riotously, and tumultuously assembled, or twelve or more of them, after proclamation made in manner aforesaid, shall continue together and not disperse themselves within one hour, That then it shall and may be lawful to and for every justice of the peace, sheriff, or under-sheriff of the county where such assembly shall be, and also to and for every high or petty constable, and other peace-officer within such county, and also to and for every mayor, justice of the peace, sheriff, bailiff, and other head-officer, high or petty constable, and other peace-officer of any city or town corporate where such assembly shall be, and to and for such other person and persons as shall be commanded to be assisting unto any such justice of the peace, sheriff or under-sheriff, mayor, bailiff, or other head-officer aforesaid (who are hereby authorized and impowered to command all his Majesty's subjects of age and ability to be assisting to them therein) to seize and apprehend, and they are hereby required to seize and apprehend such persons so unlawfully, riotously and tumultuously continuing together after proclamation made, as aforesaid, and forthwith to carry the persons so apprehended before one or more of his Majesty's justices of the peace of the county or place where such persons shall be so apprehended, in order to their being proceeded against for such their offences according to law; and that if the persons so unlawfully, riotously and tumultuously assembled, or

any of them, shall happen to be killed, maimed or hurt, in the dispersing, seizing or apprehending, or endeavouring to disperse, seize or apprehend them, that then every such justice of the peace, sheriff, under-sheriff, mayor, bailiff, head-officer, high or petty constable, or other peace-officer, and all and singular persons, being aiding and assisting to them, or any of them, shall be free, discharged and indemnified, as well against the King's Majesty, his heirs and successors, as against all and every other person and persons, of, for, or concerning the killing, maiming, or hurting of any such person or persons so unlawfully, riotously and tumultuously assembled, that shall happen to be so killed, maimed or hurt, as aforesaid.

IV.

And be it further enacted by the authority aforesaid, That if any persons unlawfully, riotously and tumultuously assembled together, to the disturbance of the publick peace, shall unlawfully, and with force demolish or pull down, or begin to demolish or pull down any church or chapel, or any building for religious worship certified and registered according to the statute made in the first year of the reign of the late King William and Queen Mary, intituled, An act for exempting their Majesty's protestant subjects dissenting from the church of England from the penalties of certain laws, or any dwelling-house, barn, stable, or other out-house, that then every such demolishing, or pulling down, or beginning to demolish, or pull down, shall be adjudged felony without benefit of clergy, and the offenders therein shall be adjudged felons, and shall suffer death as in case of felony, without benefit of clergy.

V.

Provided always, and be it further enacted by the authority aforesaid, That if any person or persons do, or shall, with force and arms, wilfully and knowingly oppose, obstruct, or in any manner wilfully and knowingly lett, hinder, or hurt any person or persons that shall begin to proclaim, or go to proclaim according to the proclamation hereby directed to be made, whereby such proclamation shall not be made, that then every such apposing, obstructing, letting, hindering or hurting such person or persons, so beginning or going to make such proclamation, as aforesaid, shall be adjudged felony without benefit of clergy, and the offenders therein shall be adjudged felons, and shall suffer death as in case of felony, without benefit of clergy; and that also every such person or persons so being unlawfully, riotously and tumultuously assembled, to the number of twelve, as aforesaid, or more, to whom proclamation should or ought to have been made if the same had not been hindered, as aforesaid, shall likewise, in case they or any of them, to the number of twelve or more, shall continue together, and not disperse themselves within one hour after such lett or hindrance so made, having knowledge of such lett or hindrance so made, shall be adjudged felons, and shall suffer death as in case of felony, without benefit of clergy

The Combination Act of 1800

An Act to repeal an Act, passed in the last Session of Parliament, intituled, "An Act to prevent Unlawful Combinations of Workmen"; and to substitute other provisions in lieu thereof.

I.

Whereas it is expedient to explain and amend an Act [39 Geo. III, c. 81] . . . to prevent unlawful combinations of workmen . . . be it enacted . . . that from . . . the passing of this Act, the said Act shall be repealed; and that all contracts, covenants and agreements whatsoever . . . at any time . . . heretofore made . . . between any journeymen manufacturers or other persons . . . for obtaining an advance of wages of them or any of them, or any other journeymen manufacturers or workmen, or other persons in any manufacture, trade or business, or for lessening or altering their or any of their usual hours or time of working, or for decreasing the quantity of work (save and except any contract made or to be made between any master and his journeyman or manufacturer, for or on account of the work or service of such journeyman or manufacturer with whom such contract may be made), or for preventing or hindering any person or persons from employing whomsoever he, she, or they shall think proper to employ . . . or for controlling or anyway affecting any person or persons carrying on any manufacture, trade or business, in the conduct or management thereof, shall be . . . illegal, null and void. . . .

II.
No journeyman, workman or other person shall at any time after the passing of this Act make or enter into, or be concerned in the making of or entering into any such contract, covenant or agreement, in writing or not in writing . . . and every . . . workman . . . who, after the passing of this Act, shall be guilty of any of the said offences, being thereof lawfully convicted, upon his own confession, or the oath or oaths of one or more credible witness or witnesses, before any two justices of the Peace . . . within three calendar months next after the offence shall have been committed, shall, by order of such justices, be committed to and confined in the common gaol, within his or their jurisdiction, for any time not exceeding 3 calendar months, or at the discretion of such justices shall be committed to some House of Correction within the same jurisdiction, there to remain and to be kept to hard labour for any time not exceeding 2 calendar months.

III.
Every . . . workman . . . who shall at any time after the passing of this Act enter into any combination to obtain an advance of wages, or to lessen or alter the hours or duration of the time of working, or to decrease the quantity of work, or for any other purpose contrary to this Act, or who shall, by giving money, or by persuasion, solicitation or intimidation, or any other means, wilfully and maliciously endeavour to prevent any unhired or unemployed journeyman or workman, or other person, in any manufacture, trade or business, or any other person wanting employment in such manufacture, trade or business, from hiring himself to any manufacturer or tradesman, or person conducting any manufacture, trade or business, or who shall, for the purpose of obtaining an advance of wages, or for any other purpose contrary to the provisions of this Act, wilfully and maliciously decoy, persuade, solicit, intimidate, influence or prevail, or attempt or endeavour to prevail, on any journeyman or workman, or other person hired or employed, or to be hired or employed in any such manufacture, trade or business, to quit or leave his work, service or employment, or who shall wilfully and maliciously hinder or prevent any manufacturer or tradesman, or other person, from employing in his or her manufacture, trade or business, such journeymen, workmen and other persons as he or she shall think proper, or who, being hired or employed, shall, without any just or reasonable cause, refuse to work with any other journeyman or workman employed or hired to work therein, and who shall be lawfully convicted of any of the said offences, upon his own confession, or the oath or oaths of one or more credible witness or witnesses, before any two justices of the Peace for the county . . . or place where such offence shall be committed, within 3 calendar months . . . shall, by order of such justices, be committed to . . . gaol for any time not exceeding 3 calendar months; or otherwise be committed to some House of Correction . . . for any time not exceeding 2 calendar months.

IV.
And for the more effectual suppression of all combinations amongst journeymen, workmen and other persons employed in any manufacture, trade or business, be it further enacted, that all and every persons and person whomsoever (whether employed in any such manufacture, trade or business, or not) who shall attend any meeting had or held for the purpose of making or entering into any contract, covenant or agreement, by this Act declared to be illegal, or of entering into, supporting, maintaining, continuing, or carrying on any combination for any purpose by this Act declared to be illegal, or who shall summons, give notice to, call upon, persuade, entice, solicit, or by intimidation, or any other means, endeavour to induce any journeyman, workman, or other person,

employed in any manufacture, trade or business, to attend any such meeting, or who shall collect, demand, ask, or receive any sum of money from any such journeyman, workman, or other person, for any of the purposes aforesaid, or who shall persuade, entice, solicit, or by intimidation, or any other means, endeavour to induce any such journeyman, workman or other person to enter into or be concerned in any such combination, or who shall pay any sum of money, or make or enter into any subscription or contribution, for or towards the support or encouragement of any such illegal meeting or combination, and who shall be lawfully convicted of any of the said offences, upon his own confession, or the oath or oaths of one or more credible witness or witnesses, before any two justices of the Peace . . . within 3 calendar months . . . shall . . . be committed to and confined in the common gaol . . . for any time not exceeding 3 calendar months, or otherwise be committed to some House of Correction . . . for any time not exceeding 2 calendar months. . . .

XVII.

All contracts between masters or other persons, for reducing the wages of workmen or for altering the usual hours of working, or increasing the quantity of work, shall be void, and masters convicted thereof shall forfeit £20.

XVIII.

And whereas it will be a great convenience and advantage to masters and workmen engaged in manufactures, that a cheap and summary mode be established for settling all disputes that may arise between them respecting wages and work; be it further enacted . . . that, from and after 1 August . . . 1800, in all cases that shall or may arise within . . . England, where the masters and workmen cannot agree respecting the price or prices to be paid for work actually done in any manufacture, or any injury or damage done or alleged to have been done by the workmen to the work, or respecting any delay or supposed delay on the part of the workmen in finishing, the work, or the not finishing such work in a good and workman-like manner, or according to any contract; and in all cases of dispute or difference, touching any contract or agreement for work or wages between masters and workmen in any trade or manufacture, which cannot be otherwise mutually adjusted and settled by and between them, it shall and may be, and it is hereby declared to be lawful for such masters and workmen between whom such dispute or difference shall arise . . . or either of them, to demand and have an arbitration or reference of such matter or matters in dispute; and each of them is hereby authorised and empowered forthwith to nominate and appoint an arbitrator . . . to arbitrate and determine such matter or matters in dispute as aforesaid by writing, subscribed by him in the presence of and attested by one witness . . . and to deliver the same personally to the other party . . . and to require the other party to name an arbitrator in like manner within two days after such reference to arbitration shall have been so demanded; and such arbitrators so appointed . . . are hereby authorised and required to . . . examine upon oath the parties and their witnesses . . . and forthwith to proceed to hear and determine the complaints of the parties, and the matter or matters in dispute between them; and the award to be made by such arbitrators within the time herein-after limited, shall in all cases be final and conclusive between the parties; but in case such arbitrators so appointed shall not agree to decide such matter or matters in dispute, so to be referred to them as aforesaid, and shall not make and sign their award within the space of three days after the signing of the submission to their award by both parties, that then it shall be lawful for the parties or either of them to require such arbitrators forthwith and without delay to go before and attend upon one of his Majesty's justices of the Peace acting in and for the county . . . or place where such dispute shall happen and be referred, and state to such justice the points in difference between them . . . which points . . . the said justice shall . . . hear and determine, and for that purpose . . . examine the parties and their witnesses upon oath, if he shall think fit. . . .

Acknowledgments

The authors would like to acknowledge the following individuals and institutions for their invaluable support. Nick Proctor, Kelly McFall, and Tony Crider supplied invaluable guidance in the development of *Engines of Mischief*. Trey Alsup deserves especial recognition for patiently helping us explore many ways to radically rethink the underlying game mechanisms of the Marketplace phase.

We are also deeply grateful for the support of these faculty members, who subjected themselves and their students to earlier drafts of *Engines of Mischief*: Kathy Donohue, and Rachel Ferguson, Kurt Kemper, Kasee Laster, John Moser, Allison Sandman, Judy Walden, and Jace Weaver. Their patience and enthusiasm testify to the social character of innovation.

The following universities and foundations offered generous financial support to run faculty playtests: the Reacting to the Past Consortium, Temple University, and Lindenwood University. We are also grateful for institutional support from Central Connecticut State University, Elon University, and the University of Providence.

Finally, the authors would like to gratefully acknowledge their families for their unfailing love and support on the long and winding road of academic game development.

Notes

CHAPTER 1

1. From John Harland, *Ballads and Songs of Lancashire, Ancient and Modern* (London: Routledge, 1875), 193–95.

2. *A New System of Practical Domestic Economy; Founded on Modern Discoveries and the Private Communications of Persons of Experience*, 3rd ed. (London, 1823), 11–13.

3. *A New System of Practical Domestic Economy*, 60–61.

CHAPTER 2

1. Benjamin Disraeli, *Coningsby, or the New Generation*, vol. 2 (London: Henry Colburn, 1844), 2.

2. William Radcliffe, *Origin of the New System of Manufacture, Commonly Called "Power-Loom Weaving," and the Purposes for Which This System Was Invented and Brought Into Use Fully Explained In a Narrative, Containing William Radcliffe's Struggles Through Life to Remove the Cause Which Has Brought This Country to Its Present Crisis* (Stockport: J. Lomax, 1828), 63.

3. Radcliffe, *Origin of the New System of Manufacture*, 65–66.

4. Louis Simond, *Journal of a Tour and Residence in Great Britain, During the Years 1810 and 1811, by a French Traveller* (Edinburgh: Constable, 1815), 1:283.

5. Simond, *Journal of a Tour and Residence in Great Britain*, 278.

6. Simond, *Journal of a Tour and Residence in Great Britain*, 284.

7. Samuel Heinrich Spiker, *Travels through England, Wales & Scotland in the Year 1816* (London, 1820), 1:235–36.

8. Spiker, *Travels through England, Wales & Scotland*, 291.

9. Richard Guest, *A Compendious History of the Cotton-Manufacture* (Manchester, UK: J. Pratt, 1823), 4.

10. Guest, *A Compendious History of the Cotton-Manufacture*, 4.

11. Robert Southey, *Letters From England by Don Manuel Alvarez Espriella, translated from the Spanish*, 2nd ed. (London: Longman, 1808), 2:87–88.

12. Southey, *Letters From England*, 2:88–89.

13. "Jerusalem," in William Blake, Edwin John Ellis, and William Butler Yeats, *The Works of William Blake* (New York: AMS, 1893), 389.

14. William Marshall, *The Review and Abstract of the County Reports to the Board of Agriculture; from the Several Agricultural Departments of England* (York, UK, 1818), 1: 257.

15. George Gordon Byron Baron Byron and Thomas Moore, *Life, Letters, and Journals of Lord Byron* (London: J. Murray, 1839), 331; Percy Bysshe Shelley, *The Masque of Anarchy: A Poem* (London: Edward Moxon, 1832), 25.

16. Mary Wollstonecraft Shelley, *Frankenstein, or the Modern Prometheus* (London: Gibbons, 1897), 196–97.

17. Great Britain, Select Committees on Hand-Loom Weavers' Petitions, 1834–1835, *Analysis of the Evidence Taken Before the Select Committees On Hand-Loom Weavers' Petitions (1834–1835): Ordered, by the House of Commons, to Be Printed, 10 August 1835* (London, 1835), 1.

18. Great Britain, Select Committees on Hand-Loom Weavers' Petitions, *Analysis of the Evidence*, 1.

19. Great Britain, Select Committees on Hand-Loom Weavers' Petitions, *Analysis of the Evidence*, 5.

20. E. P. Thompson, *The Making of the English Working Class* (New York: Vintage, 1966), 285.

21. Malcolm Chase, *Early Trade Unionism: Fraternity, Skill and the Politics of Labour* (Milton Park, UK: Taylor & Francis, 2017), 23, 107; see also 3–55.

22. William Marshall, *The Review and Abstract of the County Reports to the Board of Agriculture; from the Several Agricultural Departments of England* (York, UK: Longman, 1818), 1:258.

23. Marshall, *The Review and Abstract of the County Reports*, 1:257–59.

24. Marshall, *The Review and Abstract of the County Reports*, 1:260.

25. Marshall, *The Review and Abstract of the County Reports*, 1:257.

26. Marshall, *The Review and Abstract of the County Reports*, 1:280–81.

27. Marshall, *The Review and Abstract of the County Reports*, 1:280–81.

28. Emma Griffin, *Liberty's Dawn: A People's History of*

the Industrial Revolution (New Haven, CT: Yale University Press, 2013).

29. "Fifth Report from Select Committee on Artizans and Machinery," in *Parliamentary Papers* (London: HMSO, 1824), 397.

30. Alexis de Tocqueville, *Journeys to England and Ireland*, translated by J. P. Mayer (New Haven, CT: Yale University Press, 1958), 107-8.

31. Chase, *Early Trade Unionism*, 54.

32. Kevin Binfield, *Writings of the Luddites* (Baltimore: Johns Hopkins University Press, 2004), 202-3.

33. Binfield, *Writings of the Luddites*, 25. A "two course Hole" was a cheap but inferior knitting practice using a mechanized frame. Binfield, *Writings of the Luddites*, 21.

34. Binfield, *Writings of the Luddites*, 98-99.

35. Chase, *Early Trade Unionism*, 2, 27, 59, 71-90, 99.

36. Chase, *Early Trade Unionism*, 82-91.

37. Chase, *Early Trade Unionism*, 82.

38. Great Britain, Select Committees on Hand-Loom Weavers' Petitions, 1834-1835, *Analysis of the Evidence Taken Before the Select Committees On Hand-Loom Weavers' Petitions (1834-1835): Ordered, by the House of Commons, to Be Printed, 10 August 1835* (London, 1835), 24.

39. In England at the time there were two types of land tenure: freehold and copyhold. Freehold land was owned outright and the owner could dispose of it as he wished before or after death. In a copyhold tenure the land was owned outright by another person and the occupier of the land only had possession of it for a certain amount of time. This time could be very long, for example three lives.

40. E. P. Thompson, *The Making of the English Working Class* (New York: Vintage, 1966), 603.

41. Richard Guest, *A Compendious History of the Cotton-Manufacture* (Manchester, UK: J. Pratt, 1823), 37-38.

42. "A New Plan for the Conduct of Reformers," *Gorgon*, no. 10 (25 July 1818): 77.

43. Samuel Bamford, *Bamford's Passages in the Life of a Radical, and Early Days* (London, 1893), 141-42.

44. "Report of the Secret Committee of the House of Commons, Respecting Certain Meetings and Dangerous Combinations," *Hansard's Parliamentary Debates*, vol. 35 (25 January-28 April 1817) (London: TC Hansard, 1817), 438.

45. "Report of the Secret Committee," 447.

46. Great Britain, Select Committees on Hand-Loom Weavers' Petitions, *Analysis of the Evidence*, 21.

47. *Gorgon*, no. 11 (1 August 1818): 81-87.

48. Great Britain, Select Committees on Hand-Loom Weavers' Petitions, *Analysis of the Evidence*, 21.

CHAPTER 4

1. E. P. Thompson, *The Making of the English Working Class* (New York: Vintage, 1966), 42.

Selected Bibliography

GENERAL HISTORIES

Clark, J. C. D. *English Society, 1688–1832: Ideology, Social Structure, and Political Practice during the Ancien Regime.* Cambridge: Cambridge University Press, 1985.

Harrison, J. F. C. *Society and Politics in England, 1780–1960.* New York: Harper & Row, 1965.

Hay, Douglas, and Nicholas Rogers. *Eighteenth-Century English Society: Shuttles and Swords.* Oxford: Oxford University Press, 1997.

Hilton, Boyd. *A Mad, Bad, and Dangerous People? England, 1783–1846.* Oxford: Oxford University Press, 2006.

Porter, Roy. *English Society in the Eighteenth Century.* New York: Penguin, 1982.

ON THE INDUSTRIAL REVOLUTION

Bottomley, Sean. *The British Patent System during the Industrial Revolution, 1700–1852: From Privilege to Property.* Cambridge: Cambridge University Press, 2014.

Chapman, S. D. "Quantity Versus Quality in the British Industrial Revolution: The Case of Printed Textiles." *Northern History* 21, no. 1 (1 January 1985): 175–92.

Griffin, Emma. *Liberty's Dawn: A People's History of the Industrial Revolution.* New Haven, CT: Yale University Press, 2013.

Hahn, Barbara. *Technology in the Industrial Revolution.* Cambridge: Cambridge University Press, 2020.

Harley, C. Knick. "Cotton Textile Prices and the Industrial Revolution." *Economic History Review* 51, no. 1 (February 1998): 49–83.

Rose, Mary B., ed. *The Lancashire Cotton Industry: A History since 1700.* Preston, UK: Lancashire County Books, 1996.

Temin, Peter. "Product Quality and Vertical Integration in the Early Cotton Textile Industry." *Journal of Economic History* 68, no. 4 (December 1988): 891–907.

Timmins, Geoffrey. *The Last Shift: The Decline of Handloom Weaving in Nineteenth-Century Lancashire.* Manchester, UK: Manchester University Press, 1993.

Williams, Eric. *Capitalism and Slavery.* Chapel Hill: University of North Carolina Press, 1994.

UPPER CLASSES

Lindert, Peter H. "Unequal English Wealth since 1670." *Journal of Political Economy* 94, no. 6 (December 1986): 1127–62.

Roberts, David. "Tory Paternalism and Social Reform in Early Victorian England." *American Historical Review* 63, no. 2 (1958): 323–37.

MIDDLE CLASSES

Fleischman, Richard K., and Lee D. Parker. "British Entrepreneurs and Pre-Industrial Revolution Evidence of Cost Management." *Accounting Review* 66, no. 2 (1991): 361–75.

Griffiths, Trevor, Philip A. Hunt, and Patrick K. O'Brien. "Inventive Activity in the British Textile Industry, 1700–1800." *Journal of Economic History* 52, no. 4 (December 1992): 881–906.

Wilson, Charles. "The Entrepreneur in the Industrial Revolution in Britain." *History* 42, no. 145 (1957): 101–17.

WORKING CLASSES

Griffin, Emma. "Diets, Hunger and Living Standards during the British Industrial Revolution." *Past and Present* 239, no. 1 (1 May 2018): 71–111.

———. *Liberty's Dawn: A People's History of the Industrial Revolution.* New Haven, CT: Yale University Press, 2013.

Hall, Robert G. "Tyranny, Work and Politics: The 1818 Strike Wave in the English Cotton District." *International Review of Social History* 34, no. 3 (December 1989): 433–70.

Hammond, J. L., and B. Hammond. *The Skilled Labourer, 1760–1832.* London: Longmans, Green, 1919.

———. *The Town Labourer, 1760–1832.* London: Longmans, Green, 1917.

Hitchcock, Tim, Pamela Sharpe, and Peter King, eds. *Chronicling Poverty: The Voices and Strategies of the English Poor, 1640–1840.* Basingstoke, UK: Palgrave Macmillan, 1997.

Pocock, J. G. A. "The Classical Theory of Deference." *American Historical Review* 81, no. 3 (1976): 516–23.

Reay, B. *Rural Englands: Laboring Lives in the Nineteenth Century.* Basingstoke, UK: Palgrave Macmillan, 2004.

Thompson, E. P. *The Making of the English Working Class*. New York: Vintage, 1966.

———. "Patrician Society, Plebeian Culture." *Journal of Social History* 7, no. 4 (1974): 382–405.

Women of the Working Class

Bush, M. L. "The Women at Peterloo: The Impact of Female Reform on the Manchester Meeting of 16 August 1819." *History* 89, no. 2 (294) (April 2004): 209–32.

Clark, A. *The Struggle for the Breeches: Gender and the Making of the British Working Class*. Berkeley: University of California Press, 1997.

Schwarzkopf, Jutta. *Unpicking Gender: The Social Construction of Gender in the Lancashire Cotton Weaving Industry, 1880–1914*. London: Routledge, 2004.

Children of the Working Class

Honeyman, Katrina. *Child Workers in England, 1780–1820: Parish Apprentices and the Making of the Early Industrial Labour Force*. London: Routledge, 2016.

Working-Class Protest and Organizations

Binfield, K. *Writings of the Luddites*. Baltimore: Johns Hopkins University Press, 2004.

Bohstedt, John. *The Politics of Provisions: Food Riots, Moral Economy, and Market Transition in England, c. 1550–1850*. London: Routledge, 2016.

Chase, Malcolm. *Early Trade Unionism: Fraternity, Skill and the Politics of Labour*. London: Taylor & Francis, 2017.

Cordery, Simon. *British Friendly Societies, 1750–1914*. Basingstoke, UK: Palgrave Macmillan, 2003.

Hall, Robert G. "Tyranny, Work and Politics: The 1818 Strike Wave in the English Cotton District." *International Review of Social History* 34, no. 3 (December 1989): 433–70.

Hobsbawm, Eric, and George Rudé. *Captain Swing: A Social History of the Great English Agricultural Uprising of 1830*. New York: W. W. Norton, 1968.

Hopkins, Eric. *Working-Class Self-Help in Nineteenth-Century England*. London: Routledge, 2018.

Merchant, Brian. *Blood in the Machine: The Origins of the Rebellion against Big Tech*. New York: Little, Brown, 2023.

Patterson, A. Temple. "Luddism, Hampden Clubs, and Trade Unions in Leicestershire." *English Historical Review* 63, no. 247 (April 1948): 170–88.

Rudé, G. F. E. *The Crowd in History: A Study of Popular Disturbances in France and England, 1730–1848*. New York: Wiley, 1964.

Sale, K. *Rebels against the Future: The Luddites and Their War on the Industrial Revolution*. Reading, MA: Basic Books, 1996.

Thompson, E. P. "The Moral Economy of the English Crowd in the Eighteenth Century." *Past and Present* 50 (February 1971): 76–136.

POOR LAWS AND CHARITY

Baugh, D. A. "The Cost of Poor Relief in South-East England, 1790–1834." *Economic History Review* 28, no. 1 (1 February 1975): 50–68.

Boyer, George R. "Malthus Was Right after All: Poor Relief and Birth Rates in Southeastern England." *Journal of Political Economy* 97, no. 1 (1989): 93–114.

Lees, Lynn Hollen. *The Solidarities of Strangers: The English Poor Laws and the People, 1700–1948*. Cambridge: Cambridge University Press, 1998.

Lloyd, Sarah. *Charity and Poverty in England, c. 1680–1820: Wild and Visionary Schemes*. Manchester, UK: Manchester University Press, 2009.

Morris, R. J. "Voluntary Societies and British Urban Elites, 1780–1850: An Analysis." *Historical Journal* 26, no. 1 (1983): 95–118.

Persky, Joseph. "Classical Family Values: Ending the Poor Laws as They Knew Them." *Journal of Economic Perspectives* 11, no. 1 (Winter 1997): 179–89.

CORN LAWS

Dorfman, Robert. "Thomas Robert Malthus and David Ricardo." *Journal of Economic Perspectives* 3, no. 3 (Summer 1989): 153–64.

Williamson, Jeffrey G. "The Impact of the Corn Laws Just Prior to Repeal." *Explorations in Economic History* 27 (1990): 123–56.

ELECTIONS AND POLITICS

Gatrell, Vic. *The Cato Street Conspiracy: A Tale of Liberty and Revolution in Regency London* Cambridge: Cambridge University Press, 2022.

O'Gorman, Frank. "Electoral Deference in 'Unreformed' England: 1760–1832." *Journal of Modern History* 56, no 3 (September 1984): 391–429.

Richards, Michael. "The Lower Classes and Politics, 1800–1850." *International Labor and Working-Class History* 12 (1977): 32–26.

www.ingramcontent.com/pod-product-compliance
Lightning Source LLC
Chambersburg PA
CBHW080636230426
43663CB00016B/2892